ChatGPT for Marketing

Learn Practical Applications of ChatGPT for Marketing

Eldar Najafov

ChatGPT for Marketing: Learn Practical Applications of ChatGPT for Marketing

Eldar Najafov
Baku, Azerbaijan

ISBN-13 (pbk): 979-8-8688-0311-6   ISBN-13 (electronic): 979-8-8688-0312-3
https://doi.org/10.1007/979-8-8688-0312-3

Managing Director, Apress Media LLC: Welmoed Spahr
Acquisitions Editor: Celestin Suresh John
Development Editor: James Markham
Editorial Project Manager: Gryffin Winkler

Cover designed by eStudioCalamar

Cover image designed by Freepik (www.freepik.com)

Distributed to the book trade worldwide by Springer Science+Business Media New York, 1 New York Plaza, Suite 4600, New York, NY 10004-1562, USA. Phone 1-800-SPRINGER, fax (201) 348-4505, e-mail orders-ny@springer-sbm.com, or visit www.springeronline.com. Apress Media, LLC is a California LLC and the sole member (owner) is Springer Science + Business Media Finance Inc (SSBM Finance Inc). SSBM Finance Inc is a **Delaware** corporation.

For information on translations, please e-mail booktranslations@springernature.com; for reprint, paperback, or audio rights, please e-mail bookpermissions@springernature.com.

Apress titles may be purchased in bulk for academic, corporate, or promotional use. eBook versions and licenses are also available for most titles. For more information, reference our Print and eBook Bulk Sales web page at http://www.apress.com/bulk-sales.

Any source code or other supplementary material referenced by the author in this book is available to readers on GitHub. For more detailed information, please visit https://www.apress.com/gp/services/source-code.

If disposing of this product, please recycle the paper

Table of Contents

TABLE OF CONTENTS

About the Author

 **Eldar Najafov** has over eight years of industry experience in web development and software engineering. Beyond his technical prowess, Eldar possesses a keen understanding of project management principles. His curiosity has led him to explore AI, specifically ChatGPT. His hands-on experience in integrating ChatGPT into applications showcases his adaptability with emerging technologies.

About the Technical Reviewer

Sumit Dhami is the co-founder and COO of 3rd Shade, an innovative advertising and consulting agency in Pune. An MBA graduate from Deakin University, he specializes in strategic planning and business development. Sumit is also an AI and ChatGPT educator and practitioner, dedicated to leveraging AI to enhance business strategies. His work empowers businesses across various sectors, promoting sustainable growth and continuous learning.

Introduction

In the digital age, where technology rapidly evolves and reshapes industries, the fusion of artificial intelligence (AI) and marketing has emerged as a transformative force. At the forefront of this revolution stands OpenAI's GPT (generative pre-trained transformer) technology, a groundbreaking innovation that has catalyzed a paradigm shift in how businesses engage with their audiences. This book, *ChatGPT for Marketing: Learn Practical Applications of ChatGPT for Marketing*, delves into the intersection of AI and marketing, exploring the myriad ways in which GPT's open architecture has revolutionized marketing strategies, consumer engagement, and brand experiences.

The evolution of marketing in the digital era has been characterized by a shift toward personalized, data-driven approaches. Traditional marketing methods, while effective in their time, have been eclipsed by the vast potential of AI to analyze vast datasets, extract insights, and deliver tailored experiences at scale. OpenAI's GPT, in particular, has garnered significant attention for its ability to generate humanlike text, enabling businesses to automate content creation, customer support, and conversational interfaces with unprecedented sophistication.

The genesis of OpenAI's GPT can be traced back to the pursuit of artificial general intelligence (AGI), the holy grail of AI research. While AGI remains an aspirational goal, the development of GPT represents a significant milestone toward achieving more advanced AI capabilities. By pre-training on diverse corpora of text data, GPT models learn to understand and generate humanlike language, exhibiting remarkable fluency, coherence, and contextual awareness. This pre-training, combined with fine-tuning on specific tasks or domains, empowers GPT to adapt to a wide range of applications, including marketing.

One of the key features that sets OpenAI's GPT apart is its open architecture, which democratizes access to state-of-the-art AI capabilities. Unlike proprietary AI solutions that are restricted to select organizations or platforms, GPT's open nature allows developers, marketers, and innovators from across the globe to leverage its power, driving innovation and creativity in unprecedented ways. This accessibility has led to a proliferation of GPT-powered applications in various domains, including marketing, where its impact has been particularly profound.

INTRODUCTION

The integration of GPT into marketing workflows has revolutionized content creation, a cornerstone of modern marketing strategies. Historically, content creation has been a labor-intensive process, requiring human writers to produce copy for advertisements, blog posts, social media updates, and other marketing collateral. With GPT, however, marketers can automate content generation, harnessing the model's ability to produce high-quality, contextually relevant text on demand. Whether it's crafting product descriptions, composing email newsletters, or generating social media captions, GPT enables marketers to streamline their workflows, freeing up time and resources for strategic initiatives.

Beyond content creation, GPT has also reshaped customer interactions and engagement strategies. Conversational AI, powered by GPT, has become increasingly prevalent in marketing applications, enabling brands to deliver personalized, natural language experiences across various touchpoints. Chatbots, virtual assistants, and voice-enabled interfaces leverage GPT's language understanding capabilities to converse with customers, answer queries, provide recommendations, and facilitate transactions in real time. This level of interactivity not only enhances the customer experience but also enables brands to gather valuable insights into consumer preferences, behaviors, and sentiment.

Moreover, GPT has empowered marketers to optimize their campaigns and strategies through data-driven insights and predictive analytics. By analyzing vast amounts of textual data, including social media conversations, customer reviews, and market trends, GPT can identify patterns, detect emerging topics, and anticipate shifts in consumer sentiment. Armed with these insights, marketers can make informed decisions, refine their targeting strategies, and tailor their messaging to resonate with their audience effectively.

In addition to its applications in content generation, customer engagement, and data analysis, GPT has also sparked innovation in creative marketing campaigns and experiences. From interactive storytelling and immersive brand experiences to personalized product recommendations and dynamic advertising content, GPT's versatility and adaptability have inspired marketers to push the boundaries of creativity and experimentation. By harnessing the power of AI, marketers can deliver memorable, impactful experiences that captivate audiences and drive meaningful engagement with their brands.

As we embark on a journey through the pages of *ChatGPT for Marketing: Learn Practical Applications of ChatGPT for Marketing*, we will explore the myriad ways in which GPT is reshaping the marketing landscape. Through real-world examples, case studies, and expert insights, we will uncover the transformative potential of GPT to revolutionize content creation, enhance customer engagement, and drive innovation in marketing strategies. Whether you're a seasoned marketer looking to stay ahead of the curve or a curious enthusiast eager to explore the cutting edge of AI technology, this book offers a comprehensive guide to unlocking the full potential of OpenAI's GPT in marketing. Join us as we embark on an exciting exploration of the future of marketing in the age of AI.

CHAPTER 1

Overview of ChatGPT

In the dynamic landscape of artificial intelligence, ChatGPT, an OpenAI innovation, has asserted its dominance, ushering in a new era of natural language processing (NLP). This sophisticated language model, underpinned by advanced deep learning algorithms, has not merely evolved but revolutionized the way we engage with language-based tasks. Its capabilities extend well beyond simple text generation; it can understand context, infer intent, and provide responses that are not only accurate but also contextually relevant and coherent. This advancement in NLP represents a significant leap in AI technology, pushing the boundaries of what machines can achieve in terms of understanding and generating human language.

As we traverse the intricacies of ChatGPT, it becomes apparent that its far-reaching implications extend beyond conventional AI applications. The versatility of ChatGPT allows it to be integrated into various industries and applications, from customer service and technical support to creative writing and educational tools. Its ability to adapt to different domains and tasks showcases its robustness and flexibility, making it a valuable asset in both professional and personal contexts.

The bedrock of ChatGPT's success lies in its ability to seamlessly generate responses mirroring the intricacies of human conversational patterns. This human-like quality sets it apart, making it an invaluable asset across a spectrum of applications. Its conversational capabilities enable more natural and engaging interactions, which can lead to better user experiences and more effective communication. This human-like interaction is achieved through extensive training on diverse datasets, allowing ChatGPT to understand and respond appropriately to a wide range of topics and conversational nuances.

Beyond its immediate application, ChatGPT has become a beacon of innovation, opening doors to enhanced human–computer interactions, creative content generation, and more. It has the potential to transform how we interact with technology, making it more intuitive and accessible. For instance, in the realm of creative content generation, ChatGPT can assist writers in brainstorming ideas, drafting articles, or even composing

1

© Eldar Najafov 2024
E. Najafov, *ChatGPT for Marketing*, https://doi.org/10.1007/979-8-8688-0312-3_1

poetry. Its ability to generate coherent and contextually relevant text makes it a powerful tool for anyone engaged in content creation.

Furthermore, ChatGPT's influence extends into areas such as education, where it can serve as a tutor or study aid, providing explanations and answering questions on a wide range of subjects. In customer service, it can handle inquiries efficiently, providing timely and accurate responses that improve customer satisfaction. The potential applications are vast and varied, illustrating the transformative impact of this technology on our daily lives and professional activities.

The emergence of ChatGPT marks a pivotal moment in the evolution of artificial intelligence. Its sophisticated language model and advanced deep learning algorithms have revolutionized natural language processing, setting a new standard for human–computer interaction. As we continue to explore its capabilities, the potential for innovation and enhancement across various fields remains boundless, solidifying ChatGPT's role as a cornerstone of modern AI technology.

Revolutionizing Natural Language Processing

At its core, ChatGPT is designed to be not just a responder but a conversational participant. It is equipped with the ability to understand and generate human-like text, which allows it to engage in meaningful and coherent conversations. This capability is powered by a sophisticated neural network architecture that has been trained on a diverse and extensive dataset, encompassing a wide array of topics and linguistic nuances. This extensive training enables ChatGPT to grasp the subtleties of language, such as idiomatic expressions, contextual cues, and emotional tones, making its interactions more natural and intuitive.

Drawing from this extensive dataset, ChatGPT's understanding transcends mere accuracy, venturing into the realm of contextual relevance across diverse domains. It is not limited to responding based on keywords or simple patterns; instead, it comprehends the context and intent behind the words, allowing it to provide responses that are pertinent and insightful. This depth of understanding makes ChatGPT a powerful tool for tasks that require nuanced communication, such as customer service, technical support, and personalized content creation.

This adaptability positions ChatGPT as a versatile tool, catering to a myriad of user needs. In professional settings, it can be employed to automate routine tasks, freeing up human resources for more complex and strategic activities. For instance, it can handle customer inquiries, draft emails, generate reports, and even assist in project management by providing timely updates and reminders. In creative domains, ChatGPT can aid in brainstorming sessions, help authors overcome writer's block, and generate engaging content for blogs, social media, and marketing campaigns.

Whether unraveling complex queries, assisting in creative content generation, or streamlining routine tasks, ChatGPT has cemented its role as a transformative force. Its ability to understand and generate language with human-like proficiency enables it to bridge the gap between human and machine interaction, fostering a more integrated and efficient workflow In educational environments, ChatGPT can serve as a tutor, offering explanations, answering questions, and providing personalized learning experiences. This application is particularly valuable in remote learning scenarios, where it can supplement human instruction and offer students immediate support.

Moreover, ChatGPT's impact extends to research and data analysis. Its capacity to process and analyze large volumes of text data swiftly and accurately makes it an invaluable asset for researchers and analysts. It can assist in summarizing research papers, extracting key information from vast datasets, and even generating hypotheses based on existing data This capability not only accelerates the research process but also enhances the accuracy and depth of analysis.

The transformative power of ChatGPT lies in its ability to adapt and evolve. As it continues to interact with users and process new information, it learns and improves, refining its responses and expanding its knowledge base. This continuous learning process ensures that ChatGPT remains relevant and effective, capable of meeting the ever-changing demands of various industries and applications.

ChatGPT in Marketing

In the dynamic field of marketing, the relevance of ChatGPT takes center stage. As we explore the symbiotic relationship between ChatGPT and marketing professionals, it becomes evident that this technology is not just a tool but a strategic ally. Its integration into daily marketing operations paves the way for a responsive, dynamic approach that aligns seamlessly with the ever-evolving demands of diverse markets.

Moreover, ChatGPT's adeptness in content generation emerges as a key asset for marketers seeking resonance in their brand communication. From crafting persuasive product descriptions to tailoring engaging social media posts, ChatGPT's linguistic finesse empowers marketers to maintain a consistent and compelling brand voice.

The intricate dance between technology and creativity finds its rhythm, offering marketers a potent tool to navigate the challenges of digital communication.

As we traverse the chapters ahead, the exploration of ChatGPT's nuanced applications in marketing will delve into the ways it revolutionizes strategies and reshapes the landscape of customer engagement. The intersection of technology and marketing is no longer a distant prospect but a vibrant reality, with ChatGPT leading the charge.

Benefits of Using ChatGPT for Daily Tasks

As we embark on a comprehensive exploration of ChatGPT's impact on marketing, it becomes evident that the multifaceted benefits extend far beyond the initial efficiencies. This technology serves as a catalyst for innovation, transforming the way marketing professionals approach their daily tasks and redefining the dynamics of customer engagement.

Enhancing Efficiency Through Automation

The cornerstone of ChatGPT's contribution lies in its ability to automate repetitive and time-consuming tasks, an attribute that significantly enhances the efficiency of marketing operations. This automation ranges from generating product descriptions to responding to customer inquiries. Traditionally, these tasks required considerable human effort and time, often leading to bottlenecks in marketing workflows. ChatGPT alleviates these issues by swiftly generating high-quality content that meets the specific requirements of various marketing channels.

For example, creating compelling product descriptions is crucial for ecommerce platforms, where detailed and persuasive descriptions can drive sales. ChatGPT can generate these descriptions at scale, ensuring consistency in tone and style across a product catalog. This capability not only saves time but also ensures that all product descriptions are optimized for search engines and tailored to attract potential customers.

In addition to product descriptions, ChatGPT excels at handling customer inquiries. It can be integrated into customer support systems to provide instant, accurate responses to common questions. This reduces the workload on human support agents, allowing them to focus on more complex and nuanced customer issues. By liberating marketing workers from mundane tasks, ChatGPT opens up a realm of possibilities, allowing them to redirect their focus toward more strategic and creative endeavors. This shift not only improves overall productivity but also fosters a work environment where creativity thrives, ultimately leading to better outcomes and increased job satisfaction.

Personalized Customer Engagement

The capacity of ChatGPT to provide consistent and personalized responses to customer inquiries represents a paradigm shift in customer engagement strategies. Unlike traditional automated systems that rely on scripted responses, ChatGPT uses natural language processing to understand the context and intent behind customer queries. This allows it to generate responses that are not only accurate but also tailored to the individual customer's needs and preferences.

For instance, when integrated with customer relationship management (CRM) systems, ChatGPT can access customer data and history to personalize interactions. If a customer has previously inquired about a particular product, ChatGPT can reference this information in future interactions, providing a seamless and personalized customer experience. This level of personalization enhances customer satisfaction, as customers feel valued and understood.

Moreover, ChatGPT's ability to handle multiple languages and cultural nuances makes it an invaluable tool for global businesses. It can engage with customers from different regions in their native languages, ensuring that the customer experience is consistent and culturally appropriate. This global reach is particularly beneficial for businesses looking to expand their markets and build strong relationships with a diverse customer base.

The result of these capabilities is not just efficiency but a qualitative improvement in the way businesses interact with their clientele. Personalized customer engagement fosters loyalty and trust, which are essential for long-term business success. Customers who receive prompt, accurate, and personalized responses are more likely to return and recommend the business to others.

Strategic and Creative Focus

By automating routine tasks and enhancing customer engagement, ChatGPT allows marketing professionals to focus on strategic and creative aspects of their work. This includes developing innovative marketing campaigns, exploring new market opportunities, and crafting compelling brand narratives. Freed from the burden of repetitive tasks, marketing teams can leverage their expertise to drive business growth and achieve competitive advantage.

The integration of ChatGPT into marketing operations also supports continuous improvement. As the technology interacts with customers and processes new data, it learns and evolves, refining its responses and expanding its capabilities. This continuous learning process ensures that the technology remains effective and relevant, capable of adapting to changing market conditions and customer preferences.

Analyzing Data

Another pivotal advantage that ChatGPT brings to the marketing arena is its ability to generate reports and analyze data swiftly and accurately. The traditional methods of manual data analysis are not only time-consuming but also prone to errors. ChatGPT's data processing capabilities significantly mitigate these challenges, providing marketing professionals with timely and error-free insights. These insights become the bedrock for informed decision-making, empowering businesses to refine their products and services based on real-time market trends.

The book weaves through the various dimensions of ChatGPT's impact on daily marketing tasks, unraveling the nuances of its benefits in a more granular fashion. Each chapter delves into specific aspects, providing practical tips and strategies for marketing professionals to maximize the utility of ChatGPT.

ChatGPT's remarkable capacity to generate substantial amounts of content is a game-changer in maintaining a dynamic and relevant online presence. Crafting detailed product descriptions and composing engaging social media posts are not mere tasks for ChatGPT; they are avenues for businesses to communicate effectively with their audience.

The technology's prowess in content creation goes beyond the rudimentary generation of text. It has the innate ability to adapt to evolving trends and consumer preferences, ensuring that the marketing materials it produces are not only consistently updated but also aligned with the ever-changing digital landscape. The speed and scale at which ChatGPT operates empower marketers to keep their online content fresh, relevant, and in sync with the pulse of their target audience.

This aspect of ChatGPT's functionality is not just about generating content; it's about curating a brand narrative that resonates with the audience. The technology becomes an integral part of the storytelling process, ensuring that the brand message is not just communicated but embraced by the audience. It transforms the act of content creation from a routine task to a strategic initiative that shapes the brand's identity in the digital realm.

Learning from the Past

Furthermore, ChatGPT's ability to learn from previous interactions contributes to the refinement of content over time. This iterative learning process ensures that the content it generates becomes increasingly accurate, relevant, and aligned with the evolving preferences of the audience. In essence, ChatGPT becomes a dynamic content creation partner, adapting to the nuances of language, style, and tone that resonate most effectively with the target demographic.

The chapters ahead in this exploration will delve into the intricacies of leveraging ChatGPT's content generation capabilities. From practical applications to advanced strategies, marketing professionals will gain insights into how this technology can elevate their content creation processes, making them not just efficient but truly impactful.

A notable dimension of ChatGPT's capabilities is its learning proficiency from previous interactions, marking a significant stride in the enhancement of customer engagement. As the technology analyzes and learns from each interaction, it progressively refines its accuracy and relevance, setting the stage for more effective and efficient customer engagement.

This iterative learning process is a key driver behind ChatGPT's ability to provide more tailored and contextually appropriate responses. Each customer interaction becomes a learning opportunity, allowing the technology to discern patterns, preferences, and trends within the vast sea of conversational data. The outcome is not just responses that answer queries but responses that resonate with the unique needs and expectations of individual customers.

The journey of ChatGPT toward enhanced customer engagement goes beyond the immediate transactional aspect. It becomes a strategic initiative to understand customer preferences, predict trends, and anticipate evolving needs. Marketing professionals can leverage this capability to not only address current customer inquiries but also to proactively shape future interactions, creating a customer-centric approach that transcends traditional boundaries.

Furthermore, ChatGPT's learning proficiency contributes to a better understanding of customer preferences and emerging trends.

From tailoring product offerings to refining communication strategies, ChatGPT's journey toward enhanced customer engagement becomes a valuable asset for businesses seeking to stay ahead in a competitive market.

Practical Applications

The subsequent chapters will delve into practical applications of ChatGPT's learning capabilities, providing marketing professionals with actionable insights on how to harness this aspect of the technology for optimal customer engagement.

Team Dynamics

The collaborative potential of ChatGPT within marketing departments stands as a testament to its transformative impact on team dynamics. By automating routine tasks, ChatGPT liberates marketing professionals from the constraints of mundane activities, paving the way for increased collaboration, innovation, and overall job satisfaction.

The automation of routine tasks isn't merely about efficiency gains; it's about unlocking the creative potential within the marketing team. Freed from the burden of repetitive tasks, marketing professionals can redirect their focus toward more strategic and creative projects. This shift from a task-centric to a project-centric approach fosters a work environment where innovation thrives, ultimately leading to better outcomes.

Moreover, ChatGPT becomes an active participant in the collaborative process. As it engages in content generation, analyzes data, and refines its responses, it becomes a dynamic element in the brainstorming sessions within the marketing team. This collaborative interplay between human creativity and artificial intelligence results in a synergy that amplifies the impact of marketing initiatives.

The impact of ChatGPT on team dynamics extends beyond the immediate projects. By fostering a collaborative work environment, it contributes to increased job satisfaction among the marketing workforce. Marketing professionals find themselves engaged in more meaningful, strategic endeavors, leading to a sense of fulfillment and accomplishment. This positive work environment becomes a driving force for employee retention and overall team cohesion.

Global Reach

The global reach of ChatGPT emerges as a pivotal advantage for businesses with an international presence. Its support for a wide range of languages transcends linguistic barriers, facilitating improved customer engagement and support on a global scale. This aspect of ChatGPT's functionality is not merely an operational feature; it's a strategic tool for businesses seeking to create a seamless experience for customers worldwide.

The ability of ChatGPT to support a wide range of languages becomes a linchpin for international customer engagement. Whether a customer communicates in English, Spanish, Mandarin, or any other language, ChatGPT ensures that the interaction is not hindered by linguistic differences. This inclusivity contributes to a positive customer experience, creating a sense of accessibility and responsiveness that transcends cultural boundaries.

Moreover, ChatGPT becomes an invaluable asset for businesses in the realm of international marketing. Crafting content that resonates with diverse cultural nuances is a complex endeavor. ChatGPT's ability to adapt to different languages, tones, and styles empowers marketers to create content that is not just translated but culturally relevant. This ensures that marketing materials are not just disseminated internationally but are embraced by diverse audiences.

The global reach of ChatGPT is not just about customer-facing interactions; it's about creating a collaborative and inclusive work environment within international marketing teams. By supporting multiple languages, ChatGPT becomes a facilitator for effective communication and collaboration, transcending geographical distances and time zones.

As the book progresses, it will delve into strategies for marketing professionals to leverage ChatGPT for international business endeavors. From language-specific content creation to cross-cultural communication tips, the exploration aims to provide actionable insights for businesses aiming to harness the global reach of ChatGPT.

As we chart the future of marketing with ChatGPT, it becomes imperative to recognize that the journey has just begun. The technology is not a static tool but a dynamic force that evolves with each interaction, each project, and each innovation. The book aims to unravel the intricacies of leveraging ChatGPT, providing marketing professionals with a road map to unlock its full potential.

The exploration encompasses not just the "how" but the "why" of integrating ChatGPT into daily marketing operations. It goes beyond the technical aspects to delve into the strategic implications, offering a holistic perspective on how ChatGPT can be a game-changer for marketing strategies in the digital age.

From the automation of routine tasks to the dynamic generation of content, from enhanced customer engagement to fostering collaboration within teams, and from global inclusivity to international business endeavors, ChatGPT represents a paradigm shift. It's not just a tool; it's a partner in the journey of redefining marketing strategies and navigating the ever-evolving landscape of customer expectations.

Real-World Use Cases of ChatGPT

Example 1: Drafting Emails

Scenario: A busy professional needs to draft multiple emails quickly.

Steps:

1. **Initial Prompt:** Provide ChatGPT with a brief description of the email's purpose and key points.

   - **Example Prompt:** "Draft an email to a client informing them about the delay in their product shipment due to unforeseen circumstances. Apologize for the inconvenience and assure them of expedited shipping once the issue is resolved."

2. **Review and Edit:** Review the draft generated by ChatGPT and make any necessary edits to personalize the message.

3. **Send Email:** Once satisfied with the content, send the email.

Outcome: The professional saves time and ensures that the communication is clear and professional.

Sample Output from ChatGPT:

Subject: Update on Your Product Shipment

Dear [Client Name],

I hope this message finds you well. I am writing to inform you about an unexpected delay in the shipment of your recent order. Due to unforeseen circumstances, we are currently unable to dispatch your product as scheduled.

Please accept our sincerest apologies for any inconvenience this may cause. We understand the importance of timely delivery and are working diligently to resolve the issue. As soon as the problem is rectified, we will expedite your shipment to ensure it reaches you as quickly as possible.

Thank you for your understanding and patience. If you have any questions or require further assistance, please do not hesitate to contact us.

Best regards

[Your Name]

[Your Position]

[Your Company]

Example 2: Content Creation for Social Media

Scenario: A social media manager needs engaging content for a new product launch.
   Steps:

1. **Initial Prompt:** Provide ChatGPT with details about the product and the desired tone for the posts.

   - **Example Prompt:** "Create a series of three social media posts to promote our new eco-friendly water bottle. Highlight its features like being made from recycled materials, keeping drinks cold for 24 hours, and having a sleek design."

2. **Generate Posts:** Review the posts generated by ChatGPT and select the ones that best fit the brand's voice.

3. **Schedule Posts:** Use a social media management tool to schedule the posts.

Outcome: The social media manager efficiently generates creative and engaging content, ensuring a consistent and compelling presence across platforms.

Sample Output from ChatGPT:

> **Post 1**: ❁ Introducing our new Eco-Friendly Water Bottle! ❁ Stay hydrated and help the planet with our bottle made from 100% recycled materials. Durable, stylish, and sustainable - it's the perfect companion for your daily adventures! 🌍 🍃 #EcoFriendly #StayHydrated #Sustainability

> **Post 2**: Keep your drinks cold for up to 24 hours with our new Eco-Friendly Water Bottle! ❄ Whether you're at the gym, office, or hiking, enjoy fresh and cold beverages all day long. ❀ #EcoFriendly #CoolDrinks #NewArrival

> **Post 3**: Sleek design meets sustainability! Our Eco-Friendly Water Bottle not only looks great but also supports a greener planet. 🌱 Carry it with pride and make a statement. 🚰 #EcoFriendly #StylishSustainability #GoGreen

Example 3: Enhancing Customer Support

Scenario: An online retailer wants to provide immediate assistance to customers visiting their website.

Steps:

Initial Setup: Integrate ChatGPT with the customer support chat system on the website.

Training: Provide ChatGPT with a database of frequently asked questions and company policies.

Real-Time Interaction: Allow ChatGPT to handle initial customer inquiries and escalate more complex issues to human agents if necessary.

Outcome: Customers receive instant responses to their queries, improving their experience and reducing the workload on human support agents.

Sample Output from ChatGPT:

> Customer Query: "Can you help me track my order?"

> ChatGPT Response: "Of course! Please provide your order number, and I will check the status for you."

Example 4: Generating Ideas for Blog Posts

Scenario: A content writer needs fresh ideas for their company blog.

    Steps:

1. **Initial Prompt:** Provide ChatGPT with a brief overview of the blog's focus and target audience.

   - **Example Prompt:** "Generate a list of blog post ideas for a tech blog that focuses on the latest trends in artificial intelligence and machine learning."

2. **Review Ideas:** Select and refine the ideas generated by ChatGPT to create a content calendar.

Outcome: The writer has a ready list of engaging topics that align with the blog's theme, ensuring a steady flow of content.

    Sample Output from ChatGPT:

1. "The Future of AI: Emerging Trends to Watch in 2024"

2. "How Machine Learning is Revolutionizing Healthcare"

3. "Top 10 AI Tools for Enhancing Productivity in 2024"

4. "The Ethical Implications of AI: Balancing Innovation and Responsibility"

5. "AI in Everyday Life: How Artificial Intelligence is Shaping Our World"

Example 5: Personal Assistant Tasks

Scenario: A busy executive needs assistance with scheduling and reminders.

    Steps:

    Initial Setup: Integrate ChatGPT with the executive's calendar and task management tools.

    Daily Prompts: Use ChatGPT to generate daily summaries and reminders.

    Example Prompt: "Create a daily summary of my meetings and tasks for tomorrow."

    Outcome: The executive receives a concise and organized overview of their schedule, helping them stay on top of their commitments.

Sample Output from ChatGPT:

Good evening! Here is your schedule for tomorrow:

- 9:00 AM: Team Meeting (Conference Room A)

- 11:00 AM: Client Call with XYZ Corp

- 1:00 PM: Lunch with Sarah at Green Bistro

- 3:00 PM: Review Q2 Financial Reports

- 4:30 PM: Project Update Meeting with Development Team

Tasks:

- Prepare presentation for Friday's meeting

- Respond to emails from marketing team

- Review and approve new project proposals

Have a great day!

Summary

In conclusion, the multifaceted benefits of ChatGPT represent not just an evolution but a revolution in the field of marketing. From efficiency gains to content creation, enhanced learning capabilities, team dynamics, and global inclusivity, ChatGPT emerges as a transformative tool that empowers marketing professionals to navigate the complexities of their industry with agility and innovation.

As the chapters unfold, marketing professionals will find themselves equipped with practical tips, actionable strategies, and a deeper understanding of how ChatGPT can be woven into the fabric of their daily operations. The journey is one of innovation, collaboration, and transformation—a journey where ChatGPT is not just a tool but a catalyst for the future of marketing.

Efficiency Gains: ChatGPT automates repetitive and time-consuming tasks, enhancing overall productivity and fostering a creative work environment.

Content Creation: The technology aids in generating substantial and relevant content, ensuring a dynamic online presence.

Enhanced Learning Capabilities: ChatGPT learns from previous interactions, improving the accuracy and relevance of responses over time.

Team Dynamics: By automating routine tasks, ChatGPT encourages collaboration and innovation within marketing teams, leading to better outcomes and job satisfaction.

Global Inclusivity: Supporting multiple languages, ChatGPT facilitates improved customer engagement and support on a global scale, making content culturally relevant and inclusive.

As we chart the future of marketing with ChatGPT, it becomes imperative to recognize that the journey has just begun. The technology is not a static tool but a dynamic force that evolves with each interaction, each project, and each innovation. The book aims to unravel the intricacies of leveraging ChatGPT, providing marketing professionals with a road map to unlock its full potential. The exploration encompasses not just the "how" but the "why" of integrating ChatGPT into daily marketing operations. It goes beyond the technical aspects to delve into the strategic implications, offering a holistic perspective on how ChatGPT can be a game-changer for marketing strategies in the digital age.

From the automation of routine tasks to the dynamic generation of content, from enhanced customer engagement to fostering collaboration within teams, and from global inclusivity to international business endeavors, ChatGPT represents a paradigm shift. It's not just a tool; it's a partner in the journey of redefining marketing strategies and navigating the ever-evolving landscape of customer expectations.

CHAPTER 2

Understanding ChatGPT

ChatGPT represents a significant milestone in the field of natural language processing (NLP), showcasing the potential of transformer-based models in understanding and generating human-like text. This breakthrough technology is built upon the foundational work of transformer neural networks, which have revolutionized the way machines interpret and produce language. Unlike traditional models that processed text sequentially, transformers leverage attention mechanisms to consider the entire context of a sentence simultaneously, resulting in more coherent and contextually accurate responses.

The robust architecture of ChatGPT, which includes hundreds of millions of parameters, allows it to capture the intricacies of human language. These parameters are fine-tuned during extensive training on a diverse dataset comprising books, articles, websites, and more. This comprehensive training regimen enables ChatGPT to understand a wide range of topics, styles, and nuances, making it an adaptable tool for various applications—from drafting emails to generating creative content and providing customer support.

One of the key strengths of ChatGPT lies in its ability to continuously learn and improve. OpenAI, the organization behind ChatGPT, regularly updates the model, incorporating user feedback and new data to enhance its performance. This iterative process ensures that ChatGPT remains relevant and effective in a rapidly evolving digital landscape.

However, as powerful as ChatGPT is, its deployment comes with important considerations. Balancing efficiency and accuracy is crucial to deliver reliable results without excessive computational costs. Moreover, ethical considerations are paramount. The model's outputs are influenced by the data it was trained on, which means it can inadvertently reflect biases present in that data. OpenAI has implemented various strategies to mitigate these issues, such as refining the training process and employing human reviewers to monitor and correct biased or inappropriate outputs.

© Eldar Najafov 2024
E. Najafov, *ChatGPT for Marketing*, https://doi.org/10.1007/979-8-8688-0312-3_2

As ChatGPT continues to evolve, the importance of ethical guidelines and transparent practices cannot be overstated. Developers and users alike must remain vigilant in ensuring that the technology is used responsibly. This includes being mindful of privacy concerns, avoiding the spread of misinformation, and ensuring that the benefits of AI are accessible to all, without reinforcing societal biases or inequalities.

A Closer Look at ChatGPT

ChatGPT is a deep learning language model that utilizes advanced algorithms to generate human-like responses to a given prompt. At its core, ChatGPT is a transformer neural network, designed to process and generate text.

The model is trained on a massive amount of text data, including books, articles, and websites, allowing it to understand and generate natural language. During training, the model is exposed to different types of language patterns and text structures, allowing it to learn and generate text that is similar to human writing.

When using ChatGPT, the model is fed a prompt, or a piece of text, and generates a response based on its understanding of the prompt and its training data. The generated response is then refined using a scoring algorithm that determines the most likely response based on its quality, relevance, and grammatical correctness.

One of the remarkable features of ChatGPT is its underlying architecture, the transformer neural network. This architecture has played a pivotal role in the success of various natural language processing (NLP) tasks. Unlike traditional recurrent neural networks (RNNs) that process input data sequentially, transformers leverage attention mechanisms to capture contextual information from the entire input sequence simultaneously. This parallelization of computation results in more efficient training and better handling of long-range dependencies in language.

Evaluating Actionable Insights in Marketing Strategies

The ability of transformers to capture and leverage contextual information makes them exceptionally well-suited for marketing strategies, particularly in understanding and engaging with customer inquiries and feedback. For instance, by utilizing the attention mechanisms of transformers, models like ChatGPT can discern the nuances in customer queries, allowing for more accurate and contextually relevant responses. This capability can be transformative in automating customer service, personalizing marketing messages, and analyzing customer sentiment.

Case Study: Enhancing Customer Engagement with Transformers

Background: A leading ecommerce company faced challenges in managing a high volume of customer inquiries during peak sales periods. Traditional customer service channels were overwhelmed, leading to delayed responses and decreased customer satisfaction.

Solution: Implementing a transformer-based chatbot powered by ChatGPT to handle common customer inquiries. The chatbot was trained on historical customer service interactions, enabling it to understand and respond to a wide range of questions about product availability, shipping times, return policies, and more.

Results: The implementation of the ChatGPT-powered chatbot led to a significant reduction in response times. Customers received instant, accurate responses to their queries, improving overall satisfaction. Additionally, the chatbot's ability to handle routine inquiries allowed human agents to focus on more complex and high-priority issues.

Further Insights: Beyond handling inquiries, the transformer model analyzed customer feedback from product reviews. By categorizing sentiments and identifying recurring themes, the marketing team gained actionable insights into customer preferences. This information was instrumental in refining marketing campaigns, leading to a 15% increase in customer engagement and a 10% boost in sales during the subsequent quarter.

Educational Resources on Unsupervised Learning

One intriguing aspect of ChatGPT's training is its unsupervised nature. Unlike supervised learning, where the model is provided with labeled examples of input–output pairs, ChatGPT learns from the raw text without explicit annotations. This unsupervised learning approach enables the model to generalize well to a wide range of prompts and tasks. However, it also poses challenges, such as the potential for the model to generate incorrect or biased responses based on the patterns present in the training data.

Unsupervised learning allows the model to identify patterns and structures within the data autonomously. This capability is particularly valuable in natural language processing (NLP) because language is inherently complex and diverse. By exposing the model to vast amounts of text data, it learns to understand context, grammar, and nuances without needing specific instructions. This method enables the model to handle diverse queries, generate coherent text, and adapt to different conversational styles.

Despite its advantages, unsupervised learning presents challenges, especially regarding the quality and reliability of the generated content. Since the model relies on patterns found in the training data, it can inadvertently perpetuate biases or inaccuracies present in that data. This issue underscores the importance of using diverse and representative datasets during training and implementing robust mechanisms to identify and mitigate bias in the model's outputs.

For readers seeking to dive deeper into the mechanics of unsupervised learning and transformer models, it is recommended to explore the following educational resources:

Deep Learning by Ian Goodfellow, Yoshua Bengio, and Aaron Courville

Neural Networks and Deep Learning by Michael Nielsen

Online courses such as the "Deep Learning Specialization" on Coursera by Andrew Ng

Addressing Bias in AI-Driven Campaigns

Despite its numerous advantages, ChatGPT is not without limitations. One notable challenge is its occasional generation of responses that may be factually incorrect, biased, or nonsensical. This stems from the model's reliance on patterns learned during training, which may include inaccuracies present in the data. Addressing this limitation requires a delicate balance between refining the model's training process, incorporating diverse and reliable data sources, and implementing effective methods for bias detection and mitigation.

Ethical considerations play a crucial role in the development and deployment of ChatGPT. The model's responses are a reflection of the data it has been exposed to, and as such, it may inadvertently amplify existing biases present in the training data. OpenAI acknowledges the importance of addressing bias and actively seeks user feedback to identify and rectify instances where the model may produce biased representation or underrepresentation of certain groups. By identifying these biases early, marketers can take steps to correct them before they influence the model's outputs.

Implementing fairness-aware algorithms that adjust for biases in the training process: Fairness-aware algorithms are designed to detect and mitigate biases during the training phase. These algorithms can adjust the importance of data points to ensure that no particular group is disproportionately favored or disfavored. Techniques such as reweighting, resampling, and fairness constraints can be applied to promote equity in the model's predictions.

Using diverse datasets that reflect a wide range of perspectives and Contexts: Ensuring that training data is diverse and representative of various demographic groups is essential for reducing bias. This involves sourcing data from different geographical regions, cultural backgrounds, and socio-economic contexts. By incorporating a wide range of perspectives, marketers can develop models that are more inclusive and less likely to generate biased outputs.

Encouraging transparency and seeking continuous feedback from users to identify and mitigate biased outputs: Transparency in AI development and deployment involves openly communicating the methods used to train models, the data sources, and the potential limitations. By being transparent, organizations can build trust with users and stakeholders. Additionally, seeking continuous feedback from users helps in identifying instances of bias in real-world applications. User feedback can be instrumental in refining models and making necessary adjustments to mitigate bias.

In addition to these practices, marketers should consider appropriate outputs.

Frameworks for Bias Mitigation

To provide a thorough understanding and actionable strategies for bias mitigation, marketers should consider robust methodologies and frameworks such as regular audits of training data to identify and correct biases. Regularly reviewing and auditing training data is crucial to identify potential biases that may have been inadvertently included. This process involves examining the data for patterns that could lead to biased outcomes, such as implementing the following strategies to further strengthen bias mitigation efforts:

Bias Detection Tools: Utilizing automated tools and software designed to detect biases in datasets and model outputs can enhance the efficiency of bias mitigation efforts. These tools can scan for biased language, discriminatory patterns, and other indicators of bias, providing actionable insights for correction.

Collaborative Review Processes: Establishing a review process that involves diverse teams can help in identifying biases that a homogeneous team might overlook. Collaborators from different backgrounds can provide unique perspectives, ensuring a more comprehensive evaluation of potential biases.

Continuous Monitoring and Iteration: Bias mitigation is not a one-time task but an ongoing process. Continuous monitoring of model performance and outcomes is essential to detect and address biases that may emerge over time. Regular updates and iterations of the model, incorporating new data and feedback, help in maintaining fairness and accuracy.

Education and Training: Providing education and training on bias awareness and mitigation for all team members involved in AI development is crucial. This ensures that everyone understands the importance of bias mitigation and is equipped with the knowledge to implement best practices.

Ethical Guidelines and Policies: Developing and adhering to ethical guidelines and policies that explicitly address bias and fairness can provide a clear framework for action. These guidelines should be integrated into the organization's overall AI strategy and be regularly reviewed and updated.

Collaboration with External Auditors: Engaging external auditors to review and assess the fairness and bias in AI models can provide an unbiased perspective. External audits can validate internal efforts and offer recommendations for improvement.

By adopting these practices, marketers can maintain brand integrity and build customer trust, ensuring that AI-driven campaigns are both effective and ethically sound. Implementing robust bias mitigation frameworks not only enhances the fairness and inclusivity of AI models but also strengthens the overall credibility and reliability of the organization's marketing efforts. This commitment to ethical AI practices ultimately contributes to more positive and equitable outcomes in the marketplace.

Features of ChatGPT

ChatGPT is a highly versatile language model that offers a range of features designed to make it an indispensable tool for marketing workers. Some of the key features of ChatGPT include

> **Natural Language Processing**: ChatGPT has been trained on vast amounts of data, allowing it to understand and generate natural language, making it an ideal tool for a wide range of applications.

Automation of Routine Tasks: ChatGPT can be used to automate routine tasks, such as product descriptions and social media posts, freeing up time for more creative and strategic projects.

Content Generation: ChatGPT can quickly generate large amounts of content, such as product descriptions and social media posts, helping to keep your marketing materials up-to-date and relevant.

Customer Engagement: ChatGPT has the ability to learn from its previous interactions, improving its accuracy and relevance over time, leading to better customer engagement.

Multiple Languages: ChatGPT supports a wide range of languages, making it an ideal tool for businesses with a global reach.

Personalized Responses: ChatGPT can generate personalized responses to customer queries, improving the customer experience and building brand loyalty.

Collaboration and Teamwork: ChatGPT can help to increase collaboration and teamwork within your marketing department, leading to better outcomes and increased job satisfaction.

These are just some of the many features of ChatGPT that make it an indispensable tool for marketing workers looking to streamline their operations and stay ahead of the competition. With its ability to automate repetitive tasks, generate personalized content, and provide valuable insights, ChatGPT is a powerful tool that should not be overlooked.

As we delve further into the capabilities of ChatGPT, it becomes increasingly apparent that its versatility extends beyond its foundational role in natural language processing. While the preceding discussion illuminated the model's architecture, training process, and ethical considerations, it is equally important to explore the practical applications that make ChatGPT an invaluable asset for marketing professionals.

One of the standout features of ChatGPT is its prowess in natural language processing (NLP). The model's exposure to extensive and diverse datasets empowers it with a deep understanding of natural language, enabling it to comprehend and generate text that closely mimics human writing. For marketing workers, this translates into a tool that can effortlessly navigate the intricacies of language, delivering content that resonates with target audiences across various platforms.

Automation of routine tasks is a game-changer in the fast-paced world of marketing. ChatGPT excels in this domain, offering marketers the ability to automate repetitive tasks like crafting product descriptions and social media posts. By offloading these routine responsibilities to ChatGPT, marketing professionals can redirect their time and energy toward more creative and strategic endeavors. This not only enhances productivity but also fosters an environment conducive to innovation and fresh ideas.

Content generation is a critical aspect of marketing, and ChatGPT emerges as a reliable ally in this realm. The model's capacity to swiftly generate substantial volumes of content, be it product descriptions or social media posts, proves instrumental in keeping marketing materials up-to-date and relevant. This dynamic content creation capability aligns seamlessly with the evolving demands of the market, ensuring that marketing strategies remain agile and responsive to changing trends.

Customer engagement is a cornerstone of successful marketing, and ChatGPT contributes significantly to this aspect. The model possesses the ability to learn from its interactions, adapting and improving its accuracy and relevance over time. This iterative learning process results in more personalized and effective responses to customer queries, ultimately enhancing the overall customer experience and fostering brand loyalty. In the competitive landscape of marketing, cultivating strong customer relationships is paramount, and ChatGPT proves to be a valuable ally in achieving this goal.

The multilingual support offered by ChatGPT further expands its utility for businesses with a global reach. Marketing campaigns targeting diverse linguistic demographics can benefit from the model's proficiency in multiple languages. This feature opens doors for international expansion and allows marketers to tailor their content to specific linguistic nuances, thereby increasing the effectiveness of their communication strategies on a global scale.

Personalized responses add another layer of sophistication to ChatGPT's capabilities in customer interactions. By generating responses that are tailored to individual customer queries, the model contributes to a more personalized and human-like interaction. This level of personalization not only enhances customer satisfaction but also plays a pivotal role in building brand loyalty. In the competitive marketing landscape, where customers seek authentic and personalized experiences, ChatGPT stands out as a tool that can elevate customer engagement to new heights.

Collaboration and teamwork within a marketing department are pivotal for achieving optimal outcomes. ChatGPT, with its ability to assist in content generation, streamline processes, and provide valuable insights, emerges as a catalyst for enhanced collaboration. By automating routine tasks and offering support in generating content, the model frees up time for team members to collaborate on strategic initiatives. This collaborative synergy can lead to more innovative campaigns, improved job satisfaction, and ultimately, greater success for the marketing team as a whole.

In summary, the multifaceted features of ChatGPT position it as an indispensable tool for marketing workers seeking to streamline operations and outpace the competition. Its ability to process natural language, automate tasks, generate content, enhance customer engagement, support multiple languages, and facilitate collaboration makes it a powerhouse in the marketing arsenal. As the marketing landscape continues to evolve, leveraging the capabilities of ChatGPT becomes not just an advantage but a strategic imperative for staying at the forefront of innovation and excellence in marketing practices.

Best Practices for Using ChatGPT

In the fast-paced world of marketing, staying ahead of the curve is crucial for success. The integration of artificial intelligence (AI) has opened up new avenues for marketers to engage with their audience, and one notable tool leading this charge is ChatGPT. Developed by OpenAI, ChatGPT is a powerful language model that can be harnessed to enhance various aspects of marketing strategies. In this comprehensive guide, we will explore the best practices for utilizing ChatGPT to its fullest potential and transforming your marketing efforts.

Understanding the Basics of ChatGPT

Before diving into the best practices, it's essential to grasp the fundamentals of ChatGPT. This language model is built on the GPT-3.5 architecture, capable of understanding and generating human-like text based on the input it receives. It operates on a vast dataset and has been fine-tuned to respond contextually and coherently to a wide range of prompts. Understanding these basics ensures that marketers can effectively integrate ChatGPT into their strategies and fully exploit its capabilities.

Tailoring Content for Target Audiences

One of the key strengths of ChatGPT lies in its ability to generate personalized content. Marketers can leverage this by tailoring their messages to specific target audiences. Whether it's crafting product descriptions, email campaigns, or social media posts, using ChatGPT to create content that resonates with your audience enhances engagement and builds a stronger connection. By inputting detailed demographic information and audience preferences, marketers can ensure that the generated content aligns perfectly with the audience's interests and needs, increasing the effectiveness of their campaigns.

Enhancing Customer Support with ChatGPT

Integrating ChatGPT into customer support processes can significantly improve response times and user satisfaction. Creating chatbots powered by ChatGPT allows businesses to provide instant, round-the-clock assistance. Ensure that the responses generated by ChatGPT are regularly reviewed and updated to maintain accuracy and relevance. Additionally, ChatGPT can be trained to handle more complex queries by accessing a comprehensive knowledge base, thereby enhancing the overall customer experience and freeing up human agents to focus on more critical issues.

Streamlining Content Creation

Generating high-quality and relevant content is a time-consuming task for marketers. ChatGPT can be a valuable asset in streamlining content creation processes. By using it to draft outlines, brainstorm ideas, or even generate entire articles, marketers can save time and allocate resources more efficiently. This enables the marketing team to focus on strategic planning and creative tasks, enhancing overall productivity. Moreover, ChatGPT can assist in maintaining a consistent tone and style across various content pieces, ensuring brand coherence.

Optimizing SEO with ChatGPT

Search engine optimization (SEO) is a critical component of any digital marketing strategy. ChatGPT can be employed to create SEO-friendly content by integrating relevant keywords seamlessly. Additionally, using ChatGPT to generate meta descriptions and title tags can enhance the visibility of your content on search engine results pages. By analyzing current SEO trends and competitor strategies, ChatGPT can suggest keyword variations and content structures that are more likely to rank higher, thus driving more organic traffic to your website.

Conducting A/B Testing for Messaging

A/B testing is a proven method for refining marketing messages, and ChatGPT can play a role in this process. By generating different versions of messaging through the model, marketers can analyze which resonates best with their audience. This iterative approach ensures that marketing messages are continually optimized for maximum impact. ChatGPT can quickly generate multiple variations of a message, allowing marketers to test and implement the most effective version faster, thereby improving conversion rates and campaign performance.

Personalizing Email Campaigns

Email marketing remains a potent tool in the marketer's arsenal, and personalization is key to its success. ChatGPT can be utilized to draft personalized email content, addressing recipients by name and tailoring the message based on their preferences. This personal touch enhances engagement and increases the likelihood of conversions. By analyzing previous interactions and purchase history, ChatGPT can generate highly targeted email content that speaks directly to the recipient's interests and needs, fostering a deeper connection and loyalty to the brand.

Maintaining Ethical AI Practices

While leveraging AI models like ChatGPT, it is crucial to uphold ethical practices. Avoid using the technology to manipulate or deceive users. Be transparent about the involvement of AI in content creation, especially when interacting with customers. This ethical approach builds trust and credibility with your audience.

Monitoring and Analyzing Performance

Regularly monitoring and analyzing the performance of ChatGPT-generated content is essential for refining marketing strategies. Tracking metrics such as engagement rates, conversion rates, and customer feedback provides insights into the effectiveness of AI-driven campaigns. These insights enable marketers to make data-driven decisions and continuously optimize their approach.

Creating Interactive Experiences

Engaging with audiences through interactive experiences is increasingly popular. ChatGPT can be used to develop quizzes, polls, and conversational interfaces, making brand interactions more engaging. These interactive elements not only captivate users but also provide valuable data and insights.

Harnessing Social Media Engagement

Social media is a powerful platform for brand engagement, and ChatGPT can enhance this by generating catchy captions, responding to user comments, and creating content that aligns with trending topics. Consistent and timely social media interactions help maintain a strong online presence and foster meaningful connections with the audience.

Adapting to Evolving Trends

The digital marketing landscape is constantly evolving, and staying updated on industry trends is vital. ChatGPT can assist in this by analyzing vast amounts of data and generating insights about emerging trends and consumer behavior. Marketers can use these insights to adapt their strategies, ensuring they remain relevant and competitive.

Fostering Continuous Learning

To fully harness the potential of ChatGPT, fostering a culture of continuous learning within your marketing team is essential. Staying updated on advancements in AI technology, exploring new use cases, and encouraging team members to enhance their skills in working with ChatGPT ensures that your team remains at the forefront of AI-driven marketing strategies.

By understanding and implementing these best practices, marketers can effectively integrate ChatGPT into their strategies, leveraging its capabilities to create personalized content, improve customer engagement, and stay ahead of industry trends. Embracing

AI in marketing not only enhances efficiency and productivity but also opens up new possibilities for innovation and growth. As technology continues to evolve, those who adopt and adapt to AI-driven tools like ChatGPT will be well-positioned to thrive in the competitive digital landscape.

The Importance of Ethical AI in Marketing

The ethical use of AI in marketing is not just a theoretical concern but a practical imperative that impacts brand integrity, customer trust, and overall effectiveness of marketing campaigns. As AI technologies become more integrated into marketing strategies, it is essential to adopt ethical practices that guide the responsible use of these powerful tools.

Practical Guidelines for Ethical AI Use

To ensure ethical AI practices, marketing professionals should adhere to the following guidelines:

Transparency: Clearly communicate when AI is being used to generate content or interact with customers. This builds trust and allows customers to make informed decisions.

Accuracy: Ensure that the information provided by AI tools is accurate and reliable. Regularly review and update AI-generated content to maintain high standards of accuracy.

Bias Mitigation: Actively work to identify and eliminate biases in AI-generated content. Use diverse training datasets and implement bias detection algorithms to reduce the risk of biased outputs.

Privacy: Protect customer data by adhering to data protection regulations and implementing robust security measures. Ensure that AI tools do not misuse or mishandle personal information.

Accountability: Establish clear accountability for AI-generated content. Ensure that there are mechanisms in place for users to provide feedback and report issues with AI interactions.

Ethical Marketing: Avoid using AI to create manipulative or deceptive marketing practices. Ensure that all AI-driven campaigns align with ethical marketing standards and principles.

Developing an Ethical AI Checklist

To assist marketing professionals in implementing these guidelines, an ethical AI checklist or decision-making tool can be developed. This tool can be used to evaluate the ethical implications of AI applications in daily operations and ensure compliance with best practices.

Proposed Ethical AI Checklist for Marketing Professionals

Transparency Checklist

- Is it clear to the user when they are interacting with an AI system?

- Are AI-generated content and communications labeled as such?

Accuracy Checklist

- Is the AI system providing accurate and up-to-date information?

- Are there processes in place for regularly reviewing and correcting AI outputs?

Bias Mitigation Checklist

- Are training datasets diverse and representative?

- Are there mechanisms for detecting and mitigating biases in AI outputs?

Privacy Checklist

- Are customer data protection measures in place and compliant with regulations?

- Is AI usage aligned with the company's privacy policy?

Accountability Checklist

- Is there a clear process for users to provide feedback on AI interactions?

- Are there defined responsibilities for monitoring and managing AI outputs?

Ethical Marketing Checklist

- Do AI-driven campaigns adhere to ethical marketing standards?

- Are there safeguards against manipulative or deceptive AI practices?

Implementing the Checklist

Marketing teams should integrate the ethical AI checklist into their workflow, using it to evaluate AI applications regularly. Training sessions can be conducted to ensure that all team members are familiar with the ethical guidelines and understand how to apply the checklist effectively.

Generating Creative Campaign Ideas

Brainstorming innovative campaign ideas is a creative challenge that marketers face regularly. ChatGPT can serve as a brainstorming partner, generating unique and creative concepts based on your input. This collaborative approach can lead to breakthrough ideas that set your campaigns apart from the competition. By providing a fresh perspective and a wealth of knowledge, ChatGPT can suggest new angles, themes, and strategies that might not have been considered otherwise. For example, if a marketer is looking to promote a new product, ChatGPT can help generate a variety of campaign ideas, from catchy slogans and taglines to comprehensive campaign concepts that include digital, print, and social media elements. This can save time and inspire more innovative approaches, ultimately leading to more successful marketing campaigns.

Harnessing Social Media Engagement

Social media is a dynamic platform for brand engagement, and ChatGPT can play a role in optimizing social media content. From generating catchy captions to responding to user comments, incorporating ChatGPT into social media strategies enhances the brand's online presence and fosters meaningful interactions with the audience. ChatGPT can help create a consistent brand voice across different platforms, ensuring that all communications align with the brand's personality and values. Additionally, it can analyze engagement patterns to identify the types of content that resonate most with your audience, enabling more effective content planning. By using ChatGPT to handle routine social media interactions, brands can ensure timely and relevant responses, maintaining a high level of engagement without the need for constant manual monitoring.

Adapting to Evolving Trends

The digital landscape is constantly evolving, and marketers must stay agile to remain competitive. ChatGPT can assist in staying updated on industry trends by analyzing vast amounts of data and generating insights. Marketers can use these insights to adapt their strategies and stay ahead of the curve. For instance, ChatGPT can scan news articles, social media conversations, and industry reports to identify emerging trends and shifts in consumer behavior. It can then summarize these findings and suggest how they might impact your marketing strategies. This allows marketers to be proactive rather than reactive, adapting their campaigns to align with current trends and maintaining relevance in a fast-changing market.

Integrated Approach

By integrating these capabilities, marketers can develop a comprehensive marketing strategy where:

- **Creative campaign ideas** generated by ChatGPT provide a foundation for innovative and engaging marketing efforts.

- **Social media engagement** ensures these ideas reach a wide audience and foster meaningful interactions.

- **Trend adaptation** keeps the strategy relevant and responsive to changes in the market.

Overall, leveraging ChatGPT across these interconnected areas allows for a cohesive and dynamic marketing strategy that can lead to more successful and impactful campaigns.

Monitoring and Analyzing Performance

Regularly monitoring and analyzing the performance of ChatGPT-generated content is essential for refining marketing strategies. Track metrics such as engagement rates, conversion rates, and customer feedback to understand the impact of AI-generated content on overall marketing success. Use these insights to make data-driven decisions and continuously optimize your approach. ChatGPT can assist in this process by generating detailed performance reports, highlighting key metrics and trends. It can also suggest adjustments to content or strategy based on performance data, helping marketers to iteratively improve their campaigns. This ongoing analysis ensures that marketing efforts are always aligned with business goals and market conditions.

Creating Interactive Experiences

Engaging with audiences through interactive experiences is a trend gaining momentum in marketing. ChatGPT can be employed to develop interactive content, such as quizzes, polls, and conversational interfaces. These experiences captivate users, driving higher levels of engagement and creating memorable brand interactions. For example, a chatbot powered by ChatGPT can guide users through a personalized shopping experience on a retail website, asking questions about their preferences and making tailored product recommendations. Interactive quizzes and polls can also be used to gather valuable customer insights while entertaining your audience. By integrating these interactive elements into your marketing strategy, you can create a more engaging and immersive brand experience.

Customizing User Experiences

Tailoring user experiences based on individual preferences is a cornerstone of successful marketing. ChatGPT can assist in customizing website interactions, product recommendations, and user interfaces. By understanding user behavior and preferences, marketers can create a more personalized and enjoyable journey for their audience. For instance, ChatGPT can analyze past interactions and purchase history to provide personalized product recommendations, create dynamic website content that adapts to user preferences, and even customize email marketing campaigns to address individual interests and needs. This level of personalization helps in building stronger relationships with customers, increasing engagement, and driving conversions.

Ensuring Data Security and Privacy

To protect user data effectively, it is essential to establish a comprehensive security framework that encompasses various layers of protection. This involves implementing encryption methods to safeguard data both in transit and at rest, ensuring that any data exchanged between users and the ChatGPT system remains secure from interception or unauthorized access.

Access controls are another critical aspect of data security. By enforcing stringent access control measures, only authorized personnel are granted permission to view or manipulate sensitive information. This minimizes the risk of internal data breaches and ensures that user data is handled by qualified and trusted individuals.

Regular security audits and vulnerability assessments are vital for maintaining a strong security posture. These audits help identify potential weaknesses in the system, allowing for timely remediation before any security issues can be exploited. Keeping security measures up-to-date with the latest industry standards and best practices is essential for defending against evolving threats.

Data anonymization techniques play a significant role in protecting user privacy. By removing personally identifiable information (PII) from datasets used for training and analysis, the risk of exposing sensitive user information is greatly reduced. This approach not only protects privacy but also complies with various data protection regulations that mandate the safeguarding of PII.

Evaluating Data Security and Compliance

Given the sensitive nature of customer data used in marketing, it is crucial to ensure that robust security measures and compliance requirements are thoroughly addressed. Protecting customer data not only safeguards privacy but also maintains the integrity and trustworthiness of marketing operations. This involves implementing comprehensive data encryption protocols, stringent access controls, and regular security audits to identify and mitigate potential vulnerabilities. Additionally, staying compliant with data protection regulations such as the General Data Protection Regulation (GDPR), California Consumer Privacy Act (CCPA), and other relevant laws is essential to avoid legal repercussions and build customer trust. Transparent data handling practices and regular updates to security policies further reinforce the commitment to protecting customer data and upholding the highest standards of data security and compliance.

Detailed Strategies for Data Security and Compliance

To effectively protect data and comply with regulations, marketing professionals should implement the following strategies:

Data Encryption: Ensure that all customer data is encrypted both in transit and at rest. Encryption protects data from unauthorized access and breaches.

Access Controls: Implement strict access controls to limit who can view or manipulate customer data. Use role-based access control (RBAC) to ensure that only authorized personnel have access to sensitive information.

Data Anonymization: When using customer data for analysis or training AI models, anonymize the data to remove personally identifiable information (PII). This reduces the risk of privacy breaches.

Compliance with Regulations: Stay updated on and comply with relevant data protection regulations, such as the General Data Protection Regulation (GDPR), California Consumer Privacy Act (CCPA), and others. Implement policies and procedures that align with these regulations.

Regular Audits and Assessments: Conduct regular security audits and risk assessments to identify and address vulnerabilities. This proactive approach helps maintain a robust security posture.

Data Minimization: Collect and retain only the data necessary for your marketing operations. Limiting data collection reduces the risk of exposure in the event of a breach.

Incident Response Plan: Develop and implement an incident response plan to quickly address data breaches or security incidents. Ensure that all team members are aware of their roles and responsibilities in the event of an incident.

Industry Examples of Data Security Practices

To illustrate these strategies, here are examples from industry practices and my own agency's experience:

Example 1: Encryption and Access Controls

Scenario: An ecommerce company implemented ChatGPT to assist with customer support, which required handling customer queries and order information.

Approach: The company used end-to-end encryption for all customer data transmitted between the ChatGPT system and their servers. Additionally, they implemented role-based access control (RBAC) to ensure that only authorized support staff could access customer data.

Outcome: These measures significantly reduced the risk of unauthorized access and ensured compliance with data protection regulations.

Example 2: Data Anonymization and Compliance

Scenario: A digital marketing agency used customer feedback and interaction data to train an AI model for personalized marketing campaigns.

Approach: The agency anonymized all customer data before using it for training, removing any personally identifiable information (PII). They also implemented policies to ensure compliance with GDPR and regularly updated their procedures to reflect changes in regulations.

Outcome: This approach not only protected customer privacy but also built trust with clients by demonstrating a commitment to data security and regulatory compliance.

Example 3: Regular Audits and Incident Response

Scenario: A financial services company integrated AI-powered chatbots to handle customer inquiries and transaction information.

Approach: The company conducted regular security audits and risk assessments to identify potential vulnerabilities. They also developed a comprehensive incident response plan and trained their staff to respond effectively to data breaches.

Outcome: The proactive measures helped the company maintain a strong security posture and ensured quick and effective responses to any security incidents, minimizing potential damage.

Ensuring data security and privacy is a critical aspect of leveraging AI in marketing. By implementing detailed strategies and adhering to compliance requirements, marketing professionals can protect sensitive customer data, maintain trust, and uphold the integrity of their operations. Including these practical guidelines and examples in the book will provide readers with actionable insights to effectively secure data and comply with regulations, enhancing the overall effectiveness and ethical standards of AI-driven marketing practices.

Adapting Tone and Style

Maintaining a consistent brand voice is crucial for effective communication. ChatGPT can be trained to adopt specific tones and styles that align with your brand's personality. Whether it's a formal tone for professional communications or a more casual style for social media posts, customizing ChatGPT's output ensures a cohesive brand image.

Consistency Across Platforms

Consistency in tone and style across all communication platforms helps in building a recognizable and trustworthy brand identity. ChatGPT can be fine-tuned to reflect the desired tone across various channels, such as emails, social media, website content, and customer support interactions. This ensures that whether a customer is reading a blog post or receiving a customer service email, the messaging feels unified and consistent.

Customizing for Different Audiences

Different segments of your audience may respond better to different tones. For instance, younger audiences might prefer a more casual and engaging style, while B2B clients might appreciate a professional and formal tone. ChatGPT can be customized to adjust its tone based on the target audience, making communications more effective and resonant with each demographic.

Reflecting Brand Values

A brand's tone and style should reflect its core values and mission. Whether your brand is innovative and forward-thinking, traditional and reliable, or fun and approachable, ChatGPT can be tailored to echo these attributes. This alignment helps reinforce your brand's identity and ensures that all communications are in sync with your overall brand strategy.

Adapting to Context

The ability to adapt tone and style based on context is another key advantage of using ChatGPT. For example, while a light-hearted and friendly tone might be appropriate for social media engagement, a more serious and empathetic tone might be necessary for customer support interactions involving complaints or issues. ChatGPT can be trained to recognize these contexts and adjust its tone accordingly.

Ensuring Cultural Sensitivity

When communicating with a global audience, cultural sensitivity is crucial. ChatGPT can be programmed to be aware of cultural nuances and avoid language or tone that might be inappropriate or offensive in different cultural contexts. This capability ensures that your brand's communications are respectful and effective across diverse markets.

Regular Review and Updates

Language and communication trends evolve over time. Regularly reviewing and updating the tone and style settings in ChatGPT ensures that your brand stays current and relevant. By continuously refining the model based on customer feedback and changes in the market, you can maintain a dynamic and engaging brand voice.

Case Studies and Examples

Real-world examples of how other brands have successfully used AI to adapt their tone and style can provide valuable insights. Case studies showcasing the implementation and benefits of using ChatGPT for tone customization can serve as a guide for developing your strategy. These examples can illustrate the impact of a consistent and well-adapted brand voice on customer engagement and loyalty.

Collaboration with Human Writers

While ChatGPT can significantly enhance consistency and efficiency in brand communication, collaborating with human writers ensures a nuanced and authentic voice. Human oversight can help fine-tune the model's outputs, making adjustments that reflect subtle brand nuances and creative flair that AI might miss.

By leveraging ChatGPT to adapt tone and style, brands can achieve a cohesive, resonant, and culturally sensitive communication strategy that enhances customer experience and reinforces brand identity. This approach not only improves engagement but also builds a stronger, more reliable brand presence in the market.

Integrating Multilingual Capabilities

For global marketing efforts, incorporating multilingual capabilities into ChatGPT can be a game-changer. The model can be fine-tuned to understand and generate content in multiple languages, allowing marketers to reach diverse audiences seamlessly. This inclusivity enhances the brand's global appeal and market penetration.

Expanding Global Reach

Integrating multilingual capabilities into ChatGPT enables businesses to expand their reach to non-English-speaking markets. By communicating in the native languages of potential customers, brands can engage more effectively and build stronger connections. This capability is crucial for global brands looking to establish a presence in new regions and cater to a diverse customer base.

Cultural Relevance and Sensitivity

Beyond just translating text, it's important for ChatGPT to understand cultural nuances and context. Multilingual models can be trained to adapt content for cultural relevance, ensuring that messages resonate with local audiences. This sensitivity helps avoid misunderstandings and ensures that the brand's messaging is appropriate and effective in different cultural settings.

Enhanced Customer Support

Providing customer support in multiple languages can significantly improve customer satisfaction and loyalty. ChatGPT can handle customer inquiries, troubleshoot issues, and provide information in the customer's preferred language, ensuring a smooth and satisfactory support experience. This capability not only enhances the customer experience but also reduces the burden on human support teams by handling routine queries in various languages.

Consistency Across Languages

Maintaining a consistent brand voice across different languages is a challenge that multilingual ChatGPT can address. By fine-tuning the model to generate content that aligns with the brand's tone and style in multiple languages, marketers can ensure that the brand message remains consistent and coherent globally. This consistency helps build a strong, unified brand identity.

Localized Marketing Campaigns

Multilingual capabilities enable brands to create localized marketing campaigns tailored to specific regions. ChatGPT can generate advertisements, social media posts, email newsletters, and other marketing materials in the local language, increasing the relevance and impact of the campaigns. Localized content is more likely to resonate with the target audience, driving higher engagement and conversions.

SEO Optimization for Multiple Languages

Search engine optimization (SEO) is essential for visibility in any market. ChatGPT can assist in creating SEO-friendly content in multiple languages, incorporating relevant keywords and phrases that are commonly used in different regions. This multilingual SEO strategy helps improve the brand's search engine rankings and attract organic traffic from various linguistic markets.

Training and Customization

To maximize the effectiveness of multilingual capabilities, it's important to continuously train and customize the model. Regular updates and fine-tuning based on user feedback and market trends ensure that the model stays accurate and relevant. Additionally, incorporating feedback from native speakers can help refine the language generation to better capture idiomatic expressions and cultural subtleties.

Overcoming Language Barriers in Internal Communication

Multilingual ChatGPT can also be a valuable tool for internal communication within global organizations. It can assist in translating internal documents, memos, and emails, facilitating smoother communication among teams from different linguistic backgrounds. This capability enhances collaboration and ensures that all team members are on the same page, regardless of language barriers.

Legal and Compliance Considerations

When integrating multilingual capabilities, it's crucial to consider legal and compliance aspects related to data privacy and language-specific regulations. Ensuring compliance with local laws and regulations regarding data protection and advertising standards is essential to avoid legal issues and build trust with customers.

Implementing a Multilingual Strategy

Developing a comprehensive multilingual strategy involves identifying key markets, understanding local preferences, and customizing the model accordingly. Brands should invest in high-quality training data and collaborate with local experts to ensure the accuracy and cultural relevance of the generated content. Regularly evaluating and refining the strategy based on performance metrics and user feedback is essential for continuous improvement.

By integrating multilingual capabilities into ChatGPT, brands can effectively communicate with diverse audiences, enhance customer experiences, and strengthen their global presence. This approach not only broadens market reach but also fosters inclusivity and cultural sensitivity, positioning the brand as a truly global entity.

Adhering to Regulatory Compliance

In the ever-evolving landscape of digital marketing, staying compliant with regulations is non-negotiable. Ensure that the use of ChatGPT aligns with data protection laws, intellectual property regulations, and other relevant legal frameworks. A thorough understanding of regulatory compliance safeguards your brand from legal complications.

Comprehensive Coverage of Regulations

Given the international nature of digital marketing, it's crucial to address both global and local regulations that affect AI usage in marketing. Key regulations to consider include

> **General Data Protection Regulation (GDPR):** Applicable to any business handling the data of EU citizens, GDPR mandates strict guidelines on data collection, processing, and storage, emphasizing user consent and data protection.

California Consumer Privacy Act (CCPA): This regulation grants California residents rights over their personal data, including the right to know what data is collected and the right to request its deletion.

Personal Data Protection Act (PDPA): Relevant in Singapore, PDPA governs the collection, use, and disclosure of personal data to protect individual privacy.

Brazil's General Data Protection Law (LGPD): Similar to GDPR, LGPD regulates the processing of personal data within Brazil, emphasizing the protection of privacy and individual freedoms.

Data Security Law of the People's Republic of China (DSL): This law regulates data processing activities and aims to protect personal information, ensuring security and promoting the lawful and rational use of data.

Staying Current and Relevant

To ensure that the information remains current and relevant to global readers, regularly update the book with the latest regulatory changes and developments. Encourage businesses to stay informed about new and emerging regulations that could impact their operations.

Anticipated Changes in the Regulatory Landscape

As the digital marketing landscape evolves, so do the regulatory frameworks governing data protection and AI use. It is essential for businesses to anticipate and prepare for these changes. Consider suggesting a dedicated section on anticipated regulatory shifts and how businesses can stay ahead of these changes. This could include

Monitoring Legislative Developments: Regularly track changes in data protection laws and AI regulations in key markets. Subscribe to legal updates and participate in industry forums to stay informed.

Implementing Flexible Compliance Strategies: Develop compliance strategies that can adapt to new regulations. This includes establishing a compliance team and creating policies that can be quickly updated as laws change.

Investing in Compliance Technology: Utilize tools and technologies designed to monitor and ensure compliance with various regulations. These tools can help automate compliance tasks, reducing the risk of noncompliance.

Training and Education: Continuously educate and train employees on the importance of regulatory compliance and the specific requirements of relevant laws. Regular training sessions ensure that staff are aware of their responsibilities and the latest legal developments.

Engaging Legal Expertise: Work with legal experts who specialize in data protection and AI regulations to ensure that your business practices are compliant with current and upcoming laws.

By addressing both current regulations and anticipating future changes, businesses can effectively navigate the complex regulatory landscape of AI in marketing. This proactive approach not only ensures compliance but also builds trust with customers and stakeholders. Including these comprehensive strategies and insights in the book will provide readers with the tools and knowledge they need to manage regulatory challenges and maintain ethical AI practices in their marketing efforts.

Fostering Continuous Learning

To fully harness the potential of ChatGPT, fostering a culture of continuous learning within your marketing team is essential. Stay updated on advancements in AI technology, explore new use cases, and encourage team members to enhance their skills in working with ChatGPT. This commitment to ongoing education ensures that your team remains at the forefront of AI-driven marketing strategies.

Regular Training and Workshops

Organize regular training sessions and workshops to keep the team abreast of the latest developments in AI and machine learning. These sessions can cover new features and capabilities of ChatGPT, best practices for its application in marketing, and case studies of successful implementations. Bringing in industry experts and AI specialists can provide deeper insights and practical knowledge.

Encouraging Experimentation

Promote a culture of experimentation where team members are encouraged to test new ideas and approaches using ChatGPT. By experimenting with different use cases, from content creation to customer engagement, the team can discover innovative ways to leverage the technology. Providing a sandbox environment for experimentation allows team members to explore without the fear of making mistakes.

Knowledge Sharing

Facilitate knowledge sharing within the team by creating forums or regular meetings where team members can share their experiences, challenges, and successes with ChatGPT. This collaborative approach helps in disseminating best practices and learning from each other's experiences. Creating a repository of resources, such as tutorials, guides, and documented use cases, can also be beneficial.

Continuous Skill Enhancement

Encourage team members to pursue certifications and courses related to AI and machine learning. Platforms like Coursera, edX, and Udacity offer specialized courses that can enhance their understanding and skills in AI technologies. Supporting their participation in conferences, webinars, and industry events also helps in keeping them updated with the latest trends and advancements.

Feedback and Improvement

Implement a feedback loop where team members can provide input on their experiences with ChatGPT. This feedback can be used to identify areas for improvement and to refine the training and usage of the model. Regularly reviewing and incorporating feedback ensures that the application of ChatGPT continues to evolve and improve.

Integrating AI Ethics

Part of continuous learning involves understanding the ethical implications of using AI. Educate your team on the importance of ethical AI practices, including issues related to bias, data privacy, and transparency. By fostering an awareness of these ethical considerations, the team can make informed decisions that align with the organization's values and societal expectations.

Staying Updated with Industry Trends

The field of AI is rapidly evolving, with new advancements and trends emerging regularly. Encourage your team to stay updated by following reputable AI research publications, blogs, and news outlets. Subscribing to industry newsletters and joining professional AI and marketing associations can provide valuable insights and networking opportunities.

Leveraging AI Tools and Platforms

In addition to ChatGPT, explore other AI tools and platforms that can complement and enhance your marketing strategies. Familiarize the team with a range of AI-driven solutions for analytics, customer relationship management, and campaign automation. Understanding the broader AI ecosystem enables the team to integrate multiple technologies for more effective marketing outcomes.

Leadership Support

Ensure that the leadership team supports and prioritizes continuous learning. This includes allocating resources for training, encouraging participation in educational activities, and recognizing the efforts of team members who contribute to the learning culture. Leadership support is crucial for fostering an environment where continuous improvement is valued and pursued.

Documenting and Analyzing Results

Track and document the outcomes of using ChatGPT in various marketing initiatives. Analyzing these results helps in understanding what works well and what needs adjustment. Sharing these findings with the team contributes to collective learning and helps in refining future strategies.

By fostering a culture of continuous learning, your marketing team can fully leverage the capabilities of ChatGPT and other AI technologies. This proactive approach not only enhances individual skills and knowledge but also drives innovation and excellence in your marketing efforts. As the AI landscape continues to evolve, staying committed to ongoing education ensures that your team remains competitive and effective in delivering cutting-edge marketing solutions.

Summary

In the dynamic world of modern marketing, AI tools like ChatGPT have become essential. Developed by OpenAI, ChatGPT uses transformer-based neural networks to generate human-like text, offering versatile applications from drafting emails to customer support.

Key Features and Applications

Personalized Content: ChatGPT tailors messages for target audiences, enhancing engagement through customized product descriptions, email campaigns, and social media posts.

Customer Support: ChatGPT-powered chatbots provide instant, 24/7 assistance, improving response times and customer satisfaction.

Content Creation: It drafts outlines, brainstorms ideas, and generates full articles, saving time and resources for marketers.

SEO Optimization: ChatGPT creates SEO-friendly content by seamlessly integrating relevant keywords, boosting search engine visibility.

A/B Testing: Generates multiple versions of messaging to identify the most effective approach, optimizing marketing messages iteratively.

Interactive Experiences: Develops quizzes, polls, and conversational interfaces to engage users and gather valuable insights.

Ethical Practices and Data Security

Ethical AI Use: Emphasizes transparency, accuracy, bias mitigation, privacy, and accountability.

Data Security: Implements encryption, access controls, data anonymization, and regular audits to protect customer data and comply with regulations like GDPR and CCPA.

To maximize ChatGPT's potential, marketing teams should foster a culture of continuous learning, stay updated on AI advancements, and enhance their skills in using ChatGPT.

By integrating ChatGPT responsibly and strategically, marketers can create personalized content, improve engagement, and stay ahead of trends. Embracing AI in marketing is essential for success in today's competitive landscape, enabling marketers to innovate and connect with audiences in new, impactful ways.

CHAPTER 3

Using ChatGPT for Marketing Tasks

In this chapter, we will delve into the multifaceted applications of ChatGPT within the realm of marketing. From automating mundane tasks to crafting personalized content, ChatGPT emerges as a transformative force that can redefine your approach to marketing strategies. This advanced language model, rooted in the GPT-3.5 architecture, holds the potential to revolutionize various facets of marketing endeavors.

Integrating ChatGPT

Expanding on its applications, ChatGPT proves invaluable in the creation of interactive content, enhancement of visual elements, localized marketing efforts, and data-driven personalization. The collaboration between marketers and ChatGPT extends to strategic planning, predictive analytics, and AI-infused social listening, providing a holistic and innovative approach to marketing.

As businesses embark on this AI-powered journey, it is essential to prioritize ethical considerations. Transparency, guarding against manipulation, ensuring data security and privacy, fair and inclusive practices, and compliance with regulatory standards form the foundation for responsible and ethical ChatGPT-powered marketing.

By staying attuned to evolving trends in AI-powered marketing and fostering a culture of continuous learning within marketing teams, businesses can position themselves as pioneers in the ever-evolving digital landscape. In the intersection of technology and marketing, ChatGPT emerges not just as a tool but as a catalyst for innovation, reshaping the dynamics of how brands connect with their audiences in meaningful and impactful ways.

The integration of ChatGPT into marketing strategies offers an expansive array of possibilities for businesses seeking to enhance their outreach, engagement, and overall

© Eldar Najafov 2024
E. Najafov, *ChatGPT for Marketing*, https://doi.org/10.1007/979-8-8688-0312-3_3

success. By understanding and implementing the diverse applications of ChatGPT in content generation, customer service, market research, ad copywriting, and beyond, marketers can elevate their campaigns to new heights. The following sections take a closer look.

Content Generation

Harnessing the power of ChatGPT, marketers can streamline content generation across various channels. This includes the creation of product descriptions, blog posts, and social media content. By utilizing ChatGPT for content creation, marketers can save valuable time and resources, enabling them to redirect their efforts toward more strategic tasks. The model's ability to understand context and produce coherent, human-like text ensures the generation of high-quality content that resonates with target audiences.

Customer Service

ChatGPT serves as a valuable tool in enhancing customer service experiences. By leveraging its capabilities, businesses can provide swift and accurate responses to customer queries. This not only improves overall customer satisfaction but also fosters brand loyalty. Furthermore, the implementation of ChatGPT in customer service can alleviate the workload of support teams, allowing them to concentrate on more intricate tasks that require a human touch.

However, it is crucial to acknowledge the limitations of AI in understanding complex or nuanced queries, which may still require human intervention. While ChatGPT excels in handling routine inquiries, it may struggle with the subtleties of certain customer concerns. Therefore, integrating human oversight in the AI workflow is essential. Human agents can step in to address complex issues, ensuring that customer service remains empathetic and contextually aware.

Market Research

Incorporating ChatGPT into market research initiatives proves instrumental in obtaining valuable insights and data about target audiences. By analyzing vast datasets, the model can unveil patterns, preferences, and trends, facilitating a deeper understanding of customer needs. This data-driven approach empowers marketers to fine-tune their strategies, ensuring campaigns are precisely tailored to resonate with their audience, thereby enhancing overall campaign effectiveness.

However, the quality of insights derived from ChatGPT largely depends on the quality of the data and the robustness of the model training. Ensuring that the data used for analysis is comprehensive, up-to-date, and relevant is crucial for obtaining accurate insights. Additionally, continuously updating the model with new data helps maintain the relevance and accuracy of the insights over time. This ongoing process of data validation and model training ensures that the insights remain actionable and reflective of current market conditions.

Ad Copywriting

Crafting compelling ad copy is an art, and ChatGPT can serve as a creative collaborator in this process. By utilizing the model to generate personalized and impactful ad content, marketers can significantly enhance the performance of their advertising campaigns. Tailoring ad copy to specific target audiences ensures resonance, leading to increased conversions and sales. The adaptability of ChatGPT allows marketers to experiment with different messaging styles to optimize their approach.

Email Marketing

ChatGPT's proficiency extends to the realm of email marketing, offering the ability to generate personalized and relevant emails. By employing the model to create engaging email content, marketers can improve open rates, click-through rates, and overall conversions. The personalized touch adds a human element to email communication, fostering stronger connections with recipients and ultimately driving higher levels of engagement and sales.

Social Media Management

Effectively managing social media accounts requires consistent and engaging content, quick responses to user inquiries, and insightful analytics. ChatGPT can play a pivotal role in these areas, generating personalized content, responding to customer queries, and providing valuable insights into social media engagement. This integration enables businesses to build stronger relationships with their audience, enhancing brand presence and credibility across various social platforms.

Lead Generation

Generating leads is a fundamental aspect of marketing, and ChatGPT can contribute significantly to this process. Whether through chatbots on websites or personalized email campaigns, utilizing ChatGPT ensures the delivery of relevant and personalized information to potential leads. This personalized approach increases the likelihood of converting leads into sales, contributing to the overall success of lead generation initiatives.

Case Studies on ChatGPT-Driven Lead Generation

To illustrate the effectiveness of ChatGPT in lead generation, consider the following examples.

Case Study: Ecommerce Platform

An ecommerce company implemented ChatGPT-powered chatbots on their website to engage with visitors in real time. The chatbots provided personalized product recommendations based on user behavior and preferences. This approach resulted in a 20% increase in conversion rates and a 15% reduction in cart abandonment rates. By addressing customer queries promptly and accurately, the chatbots enhanced the overall shopping experience and drove higher sales.

Case Study: B2B Marketing Campaign

A B2B company utilized ChatGPT for a personalized email marketing campaign targeting potential clients. The AI-generated emails were tailored to the specific needs and pain points of each recipient, resulting in a 30% increase in response rates compared to previous campaigns. The campaign also saw a 25% increase in lead-to-sales conversion rates, demonstrating the impact of personalized communication in fostering stronger business relationships.

These case studies highlight best practices and lessons learned in deploying ChatGPT for lead generation. The key takeaways include the importance of understanding customer needs, utilizing data-driven insights for personalization, and continuously refining AI models to maintain effectiveness.

By incorporating ChatGPT into your marketing tasks, you unlock the potential for improved efficiency, accuracy, and relevance, ultimately leading to more favorable outcomes and increased success. Whether your goal is to automate repetitive tasks, generate personalized content, or enhance customer engagement, ChatGPT emerges as a versatile tool that can reshape the landscape of your marketing strategies.

Expanding on the Applications of ChatGPT in Marketing

To further elaborate on the transformative capabilities of ChatGPT in marketing, let's explore additional dimensions and use cases where this powerful language model can be seamlessly integrated.

Interactive Content Creation

The modern consumer craves interactive and engaging content. ChatGPT can be employed to create interactive experiences such as quizzes, polls, and surveys. These not only captivate the audience but also provide valuable data for marketers to refine their strategies. By offering interactive content, businesses can elevate user engagement, encouraging participation and building a sense of community around their brand.

Visual Content Enhancement

While ChatGPT primarily excels in text-based generation, its integration with visual content is not to be underestimated. Marketers can use the model to generate captions for images, infographics, and other visual elements. This ensures consistency in messaging across various content formats and enhances the overall appeal of visual content on platforms like social media.

Localized Marketing Efforts

For businesses operating in diverse markets, ChatGPT can be tailored to understand and generate content in multiple languages. This localization capability is crucial for crafting marketing messages that resonate with specific regional audiences. By embracing multilingual capabilities, businesses can transcend language barriers, expanding their reach and establishing a more profound connection with global consumers.

Data-Driven Personalization

In the era of data-driven marketing, personalization is key to capturing audience attention. ChatGPT can analyze user data and preferences, enabling marketers to deliver highly personalized content experiences. From suggesting tailored product recommendations to creating individualized marketing messages, ChatGPT ensures that each interaction feels bespoke, thereby fostering stronger connections between brands and consumers.

Strategic Collaboration with ChatGPT

The collaboration between marketers and ChatGPT extends beyond content creation. The model can be integrated into the strategic planning phase, helping teams generate ideas, outline content structures, and refine messaging strategies. This collaborative approach ensures a diversity of perspectives in the ideation process, leading to more innovative and effective marketing campaigns.

Enhanced Predictive Analytics

The predictive capabilities of ChatGPT can be harnessed to anticipate market trends and consumer behavior. By analyzing historical data and market indicators, the model can generate insights that guide strategic decision-making. This foresight enables marketers to proactively adjust their approaches, stay ahead of industry trends, and position their brands as trendsetters in the market.

However, it is crucial to recognize the limits of predictions based on historical data, especially in rapidly changing markets. AI predictions are inherently based on patterns observed in past data, which may not always account for sudden shifts or unprecedented events. Therefore, while ChatGPT's predictive analytics offer valuable insights, they should not be relied upon in isolation.

Combining AI predictions with human judgment is essential in strategic decision-making. Human expertise and intuition can provide context and adaptability that AI models may lack. This integrated approach ensures a more comprehensive understanding of market dynamics, allowing businesses to make informed decisions that are both data-driven and contextually aware.

By leveraging the strengths of both AI and human judgment, marketers can navigate the complexities of modern markets more effectively, ensuring that their strategies are robust, flexible, and capable of responding to unforeseen challenges.

AI-Infused Social Listening

Social listening is a vital component of understanding audience sentiment. ChatGPT can be applied to analyze social media conversations and comments, providing valuable insights into public perception. Marketers can use this information to tailor their messaging, address concerns, and capitalize on positive sentiment, ultimately shaping a more effective and responsive social media strategy.

Key Benefits of AI-Infused Social Listening:

- **Real-Time Insights**: ChatGPT can process social media data in real time, enabling marketers to respond swiftly to emerging trends or crises.

- **Comprehensive Analysis**: The model can analyze sentiment, detect sarcasm, and understand context, providing a deeper understanding of audience emotions and opinions.

- **Enhanced Customer Engagement**: By identifying what resonates with the audience, businesses can create more engaging and relevant content, fostering stronger connections with their followers.

- **Crisis Management**: Early detection of negative sentiment allows brands to manage potential PR crises before they escalate, protecting brand reputation.

Evolving Trends in AI-Powered Marketing

As technology evolves, the landscape of AI-powered marketing continues to transform. Marketers should stay abreast of emerging trends and advancements in AI technology. Regularly updating the skills of marketing teams to incorporate the latest AI tools and methodologies ensures a proactive and adaptive approach to the ever-changing dynamics of digital marketing.

Current and Emerging Trends

- **Hyper-Personalization**: AI enables highly personalized marketing campaigns that cater to individual preferences and behaviors, resulting in more effective customer engagement and higher conversion rates.

- **Voice Search Optimization**: With the rise of voice-activated assistants, optimizing content for voice search is becoming essential. AI can help tailor content to match the natural language patterns used in voice searches.

- **AI-Driven Content Creation**: Beyond text, AI is increasingly being used to generate visual content, video scripts, and even interactive media, expanding the creative possibilities for marketers.

- **Predictive Analytics**: AI's ability to analyze vast datasets and predict future trends helps marketers make data-driven decisions, improving campaign effectiveness and return on investment (ROI).

- **Chatbots and Conversational AI**: These tools are enhancing customer service and lead generation by providing instant, accurate responses and engaging users in meaningful conversations.

Ensuring Ethical Practices in ChatGPT-Powered Marketing

As businesses increasingly integrate ChatGPT into their marketing endeavors, maintaining ethical practices is paramount. Here are essential considerations to uphold ethical standards in the use of ChatGPT

Transparency in AI Integration

Be transparent with your audience about the involvement of AI, specifically ChatGPT, in content creation. Clearly communicate when AI is employed in customer interactions or content generation to build trust and transparency with your audience.

Clear Communication Strategies

- **Explicit Acknowledgment**: Clearly label content created or assisted by AI with statements like "Generated by ChatGPT" or "AI-Assisted Content." This should be prominently displayed in the content itself, ensuring that the audience can immediately recognize the involvement of AI.

- **Content Disclaimers**: For more in-depth content such as articles, blog posts, or reports, include disclaimers at the beginning or end, explaining how AI was used in the creation process. For example, "This article was generated with the assistance of ChatGPT to enhance content quality and ensure comprehensive coverage of the topic."

- **AI Usage Badges**: Incorporate visual badges or icons in your digital content that signify AI involvement. These can be placed in a consistent location, such as the footer of a web page or the signature of an email, to standardize the indication of AI use.

Enhancing Trust Through Education

- **Educational Campaigns**: Run educational campaigns to inform your audience about the benefits and limitations of AI in content creation. This can include webinars, blog posts, or social media series that explain how AI like ChatGPT works, its capabilities, and why your organization uses it.

- **Transparency Pages**: Dedicate a section of your website to transparency about AI use. This page can provide detailed information on how AI is integrated into your workflows, examples of AI-generated content, and answers to frequently asked questions about AI in your operations.

- **Interactive Demos**: Offer interactive demos or behind-the-scenes looks at how ChatGPT generates content. This can demystify the process for your audience and show the human oversight involved in using AI.

Building Trust Through Accountability

- **Feedback Mechanisms**: Establish clear feedback channels where users can report issues or provide feedback on AI-generated content. This feedback should be actively reviewed and used to improve AI applications.

- **Regular Updates**: Keep your audience informed about any updates or changes in your AI usage policies. Regularly update your transparency pages and communicate any significant changes through newsletters or social media announcements.

- **Human Oversight**: Emphasize the role of human oversight in AI processes. Make it clear that while AI assists in content creation, human editors review and refine the output to ensure accuracy and relevance.

Ethical Considerations

- **Ethical AI Guidelines**: Publish your ethical guidelines for AI use, outlining how you ensure the responsible and fair use of AI in your operations. These guidelines should cover data privacy, nonbias, and the ethical treatment of content generated by AI.

- **Bias Audits**: Conduct and publish regular bias audits of your AI systems to ensure they are producing fair and unbiased content. Share the results of these audits with your audience to demonstrate your commitment to ethical AI use.

Practical Implementation

- **Onboarding Processes**: Train your team on how to communicate AI involvement clearly and effectively. Ensure that everyone from content creators to customer service representatives understands the importance of transparency and knows how to convey it to customers.

- **Customer Service Transparency**: In customer service interactions, start with a clear indication that the initial response is generated by an AI assistant. For example, "You are chatting with our AI assistant, ChatGPT. How can I assist you today?" This sets clear expectations and builds trust from the start.

- **Content Verification**: Implement a verification process where human reviewers check AI-generated content for accuracy and relevance before it is published or shared. This human–AI collaboration ensures high-quality output and reinforces transparency.

By being transparent about the use of AI, specifically ChatGPT, in content creation and customer interactions, businesses can build trust and credibility with their audience. Clear communication, educational initiatives, and ethical practices are essential components of this transparency, ensuring that AI integration is both effective and trustworthy. This approach not only enhances customer trust but also leverages AI to its fullest potential, improving content quality and operational efficiency.

Guarding Against Manipulation

While ChatGPT can be a powerful tool, it is crucial to avoid using it to manipulate or deceive users. Uphold integrity in marketing communications and ensure that AI-generated content adheres to ethical standards. Misleading practices can damage brand reputation and erode trust among consumers.

To delve deeper into this essential ethical consideration and help marketers ensure their use of ChatGPT remains transparent and ethical, here are some guidelines and a checklist for ethical AI use in marketing.

Guidelines for Ethical AI Use in Marketing

Transparency: Always disclose when AI is used in content creation or customer interactions. Transparency builds trust and allows consumers to make informed decisions about their interactions with your brand.

User Consent: Obtain explicit consent from users before collecting and using their data. Make sure users understand how their data will be used and provide them with options to opt out.

Accuracy and Honesty: Ensure that the information generated by ChatGPT is accurate and does not mislead users. Regularly review AI-generated content for accuracy and make necessary corrections.

Data Privacy: Protect user data by implementing robust security measures. Comply with data protection regulations such as GDPR and CCPA. Avoid using personal data for purposes other than those explicitly stated to the user.

Bias and Fairness: Train AI models on diverse datasets to minimize biases. Regularly audit AI systems for bias and take corrective actions to ensure fair and inclusive content generation.

Ethical Oversight: Establish an ethical oversight committee to monitor AI usage and address potential ethical issues. This committee should include stakeholders from diverse backgrounds to ensure a holistic view of ethical considerations.

Checklist for Ethical AI Use

☐ **Disclose AI Usage**: Clearly indicate when ChatGPT is used in customer interactions or content creation.

☐ **Obtain User Consent**: Ensure all data collection and usage practices are transparent and user consent is obtained.

☐ **Verify Accuracy**: Regularly review AI-generated content for accuracy and make corrections as needed.

☐ **Protect Data Privacy**: Implement and maintain strong data security measures. Adhere to relevant data protection regulations.

☐ **Audit for Bias**: Regularly audit AI models for biases and take corrective actions to ensure fairness.

☐ **Ethical Committee**: Establish an ethical oversight committee to monitor and address ethical issues in AI use.

☐ **Provide Opt Out Options**: Allow users to opt out of AI interactions if they prefer human assistance.

☐ **Educate and Train Staff**: Provide ongoing education and training for staff on ethical AI use and data privacy.

By adhering to these guidelines and utilizing the checklist, marketers can ensure their use of ChatGPT remains ethical and transparent, fostering trust and integrity in their marketing practices.

Data Security and Privacy

Prioritize the security and privacy of user data when employing ChatGPT in marketing processes. Implement robust data protection measures to prevent unauthorized access and comply with relevant data protection regulations. Safeguarding customer information is essential for maintaining trust and credibility.

Fair and Inclusive Practices

Ensure that AI applications, including ChatGPT, are deployed with fairness and inclusivity in mind. Avoid biases in content generation and be attentive to potential ethical issues. Promote diversity and inclusivity in marketing messages, and regularly review and refine AI models to minimize unintended biases.

Implementing Fair and Inclusive AI

- **Diverse Training Data**: Utilize diverse and representative datasets to train AI models. This helps in minimizing biases and ensures that the AI understands and respects different cultures, languages, and perspectives.

- **Bias Detection Tools**: Incorporate bias detection tools and algorithms that can identify and flag biased content. Regularly audit AI outputs to ensure they meet inclusivity standards.

- **Inclusive Language:** Ensure that the language used by AI models is inclusive and respectful. Avoid terms and phrases that may be discriminatory or exclusionary. Regularly update the AI's language database to reflect evolving societal norms and values.

- **Cultural Sensitivity:** Train AI to understand and respect cultural nuances. This includes being aware of cultural holidays, significant events, and context-specific language variations to avoid cultural insensitivity in content generation.

Promoting Diversity in Marketing Messages

- **Inclusive Content Strategies:** Develop content strategies that reflect diverse voices and perspectives. Highlight stories, case studies, and testimonials from a wide range of demographics to ensure representation.

- **Collaborative Content Creation:** Involve diverse teams in the content creation process. This ensures that multiple viewpoints are considered, reducing the risk of unintentional bias.

- **Community Engagement:** Engage with diverse communities to gather insights and feedback on your content and marketing strategies. This helps in understanding different perspectives and ensuring that your messaging is inclusive and resonant with a broader audience.

Continuous Improvement and Education

- **Regular Training:** Provide ongoing training for your team on the importance of fairness and inclusivity in AI applications. Ensure that they are aware of the latest developments and best practices in ethical AI use.

- **Feedback Loops:** Establish feedback loops where customers and users can report any biases or issues they encounter with AI-generated content. Use this feedback to continuously refine and improve your AI models.

- **Transparency in Processes**: Be transparent about the steps you are taking to ensure fairness and inclusivity. Share your methods, findings, and improvements with your audience to build trust and demonstrate your commitment to ethical practices.

Compliance with Regulatory Standards

Stay vigilant about compliance with regulatory standards governing the use of AI in marketing. Familiarize yourself with data protection laws, intellectual property regulations, and other relevant legal frameworks. Adhering to these standards not only ensures legal compliance but also safeguards your brand from potential legal complications.

Understanding Regulatory Requirements

- **Data Protection Laws**: Ensure compliance with data protection regulations such as the General Data Protection Regulation (GDPR) in Europe, the California Consumer Privacy Act (CCPA) in the United States, and other relevant local laws. These laws govern how data is collected, stored, and used, and noncompliance can result in significant penalties.

- **Intellectual Property**: Respect intellectual property rights when using AI-generated content. Understand the legal implications of using content created by AI and ensure that it does not infringe on existing copyrights or trademarks.

- **AI-Specific Legislation**: Keep abreast of emerging regulations specifically targeting AI and machine learning. This includes understanding requirements for transparency, accountability, and fairness in AI applications.

Implementing Compliance Measures

- **Data Privacy Policies**: Develop and enforce robust data privacy policies that outline how customer data is handled, stored, and protected. Ensure that these policies are transparent and accessible to your customers.

- **Consent Mechanisms**: Implement clear consent mechanisms for data collection. Ensure that users are fully informed about what data is being collected and how it will be used, and provide options for users to opt out if they choose.

- **Regular Audits**: Conduct regular audits of your AI systems and data handling practices to ensure compliance with all relevant regulations. Use these audits to identify and address any potential issues proactively.

- **Legal Counsel**: Consult with legal experts who specialize in AI and data protection to ensure that your practices are compliant with current laws and regulations. This can help in navigating complex legal landscapes and avoiding potential pitfalls.

Ethical Use of AI

- **Transparency Reports**: Publish regular transparency reports detailing how AI is used within your organization, the data it processes, and the measures taken to ensure ethical use. This builds trust and shows your commitment to responsible AI use.

- **Ethical Review Boards**: Establish ethical review boards to oversee the deployment and use of AI systems. These boards should include experts from diverse fields to provide a comprehensive assessment of AI practices.

- **User Control and Rights**: Empower users with control over their data and AI interactions. Provide options for users to access, modify, and delete their data and to opt out of AI-driven services if they prefer human interaction.

Generating Product Descriptions

Product descriptions play a pivotal role in the success of any ecommerce business. These concise pieces of content serve as the bridge between your products and potential customers, offering crucial information that aids in making informed purchase decisions. However, crafting high-quality, engaging, and informative product descriptions can be a laborious and challenging task. Enter ChatGPT, a powerful tool that can revolutionize the way you create product descriptions, saving you time and enhancing the overall customer experience.

Consistent and Accurate Descriptions

One of the significant advantages of utilizing ChatGPT for product descriptions is its ability to maintain consistency and accuracy. By training the model on existing product descriptions, you ensure that the language and tone remain uniform across all your offerings. Consistency fosters a sense of reliability and professionalism, contributing to an improved user experience. Moreover, accurate descriptions build trust with your customers, as they can rely on the information provided, ultimately increasing their confidence in making a purchase.

ChatGPT's capacity to understand and replicate the nuances of your existing descriptions ensures that the essence of your brand is preserved. This feature is particularly beneficial for businesses with a diverse range of products, as it helps maintain a cohesive and recognizable brand identity throughout the product catalog.

Personalized Descriptions

Tailoring product descriptions to specific audiences and markets is a key strategy for increasing conversions and driving sales. ChatGPT excels in this area by allowing you to generate personalized product descriptions that resonate with your target audience. By inputting specific parameters and preferences, you can instruct ChatGPT to create descriptions that address the unique needs and interests of different customer segments.

Whether you're targeting different demographics, geographical locations, or customer preferences, ChatGPT enables you to produce content that speaks directly to your audience. This personalized touch enhances the relevance of your product descriptions, making them more compelling and persuasive. As a result, potential customers are more likely to connect with the product, leading to increased conversion rates and improved overall sales performance.

Faster Turnaround Time

Speed is a critical factor in the competitive landscape of ecommerce. With ChatGPT, you can significantly reduce the time it takes to generate product descriptions. Unlike human writers who may require hours or even days to produce content, ChatGPT can generate descriptions in a matter of minutes. This accelerated turnaround time is especially beneficial when introducing new products to the market or responding to sudden spikes in demand.

By leveraging ChatGPT's rapid content creation capabilities, you can stay ahead of the competition, ensuring that your products are swiftly and effectively presented to potential customers. The ability to keep pace with market trends and consumer demands contributes to maintaining your competitiveness in the dynamic ecommerce landscape.

Increased Efficiency

Using ChatGPT to handle the generation of product descriptions allows you to reallocate human resources to more strategic and creative tasks. By automating the descriptive content creation process, you free up valuable time and talent that can be directed toward other aspects of your business, such as marketing, customer service, or product development.

The increased efficiency resulting from ChatGPT's contribution to the product description workflow translates into improved overall productivity. As your team focuses on higher-level tasks, you can expect to see enhanced business outcomes, from increased innovation to better customer engagement. This efficiency gain is particularly valuable for small businesses with limited resources, allowing them to compete more effectively with larger counterparts.

In conclusion, the integration of ChatGPT into your ecommerce business for generating product descriptions can bring about significant improvements in quality, relevance, and efficiency. Whether you're a small business looking to optimize your operations or a large corporation seeking to streamline a vast product catalog, ChatGPT offers a versatile solution to meet your needs.

By ensuring consistent and accurate descriptions, tailoring content to specific audiences, reducing turnaround times, and increasing overall efficiency, ChatGPT becomes a valuable asset in the quest to enhance the customer experience. Embracing this innovative approach to product description generation positions your business for success in the competitive ecommerce landscape, providing a distinct advantage that can lead to increased sales and customer satisfaction.

Creating Email Templates

In the contemporary business landscape, email stands as a cornerstone of communication, and the efficiency of this tool is paramount. Effective email templates not only save time but also play a crucial role in maintaining consistency and professionalism in business communications. With the assistance of ChatGPT, businesses can revolutionize their email template creation process, tailoring messages to specific needs and ensuring a seamless and efficient communication flow.

Consistent and Professional Templates

One of the primary advantages of incorporating ChatGPT into your email template creation process is the ability to ensure consistency and professionalism across all communications. By training ChatGPT on existing email templates, you can maintain a standardized format, language, and tone. This consistency is not only visually appealing but also contributes to a more professional image, fostering trust with your customers.

Consistent email templates are particularly important for businesses that engage in regular email correspondence, such as customer support, newsletters, or marketing campaigns. The uniformity in communication style ensures that your brand's identity remains intact, reinforcing a sense of reliability and credibility in the eyes of your audience.

Personalized Templates

Personalization is a key driver of successful email marketing and communication. ChatGPT's ability to generate personalized email templates enables businesses to tailor their messages to specific audiences and markets. By inputting relevant information about your target audience, ChatGPT can craft email templates that resonate with the unique preferences and needs of different customer segments.

Whether you are reaching out to existing customers, leads, or different demographic groups, personalized email templates enhance the relevance of your messages. This personalized touch increases the likelihood of engagement, conversion, and ultimately, sales. The adaptability of ChatGPT in catering to diverse audiences ensures that your email communication remains impactful and customer-centric.

Faster Turnaround Time

Time is of the essence in the fast-paced world of business, and responding promptly to customer inquiries or requests is vital for maintaining a positive customer experience. ChatGPT excels in generating email templates at a much faster pace than human counterparts. This rapid turnaround time enables businesses to address customer concerns, answer inquiries, and provide information in a timely manner.

By leveraging ChatGPT's efficiency in content creation, you can significantly reduce the time it takes to draft and send emails. This is particularly beneficial for scenarios where quick responses are crucial, such as customer support, sales inquiries, or time-sensitive promotions. The ability to act swiftly contributes to your business's agility, helping you stay competitive and enhancing overall customer satisfaction.

Increased Efficiency

Using ChatGPT to generate email templates doesn't just save time; it also enhances overall efficiency by freeing up human resources for more strategic tasks. The automation of email template creation allows your team to focus on other critical aspects of your business, whether it be refining marketing strategies, improving product offerings, or enhancing customer engagement initiatives.

The increased efficiency resulting from ChatGPT's contribution to email template creation is a valuable asset for businesses of all sizes. Small businesses can allocate their limited resources more effectively, while larger corporations can optimize their workflows for maximum productivity. This efficiency gain has a cascading effect on various facets of your business, ultimately contributing to better organizational performance.

In conclusion, leveraging ChatGPT for creating email templates can be a game-changer for businesses seeking to enhance the standard, relevance, and speed of their email correspondence. The ability to ensure consistency, personalize messages, achieve faster turnaround times, and increase overall efficiency positions ChatGPT as a versatile tool in optimizing email communications.

Regardless of the size of your business, integrating ChatGPT into your email template creation process can have a profound impact on the customer's overall experience. From building trust through consistent and professional communication to driving sales through personalized messages, ChatGPT offers a multifaceted solution to elevate your email correspondence to new heights.

Automating Social Media Posts

In the ever-evolving landscape of digital marketing, social media stands as a cornerstone for brand promotion and engagement. Automating social media posts has emerged as a crucial strategy to save time, ensure consistency, and reach a broader audience with high-quality content. With the advent of ChatGPT, businesses can now seamlessly integrate automation into their social media marketing efforts, delivering a consistent brand voice and relevant and timely posts and achieving multichannel support.

Consistent Brand Voice

Maintaining a consistent brand voice across various social media platforms is essential for building brand identity and customer trust. ChatGPT can be a valuable asset in this regard, as it can be trained on your existing brand voice to ensure that all automated social media posts align with your brand's tone, language, and values.

By leveraging ChatGPT's ability to understand and replicate specific brand nuances, businesses can present a unified front to their audience. Consistency in messaging contributes to a professional image, reinforces brand recognition, and fosters a sense of trust among followers. This unified brand voice becomes even more critical as businesses expand their social media presence and engage with diverse audiences across different platforms.

Relevant and Timely Posts

One of the challenges in social media marketing is creating content that remains relevant and timely to the audience. ChatGPT excels in generating social media posts that are not only tailored to your brand voice but are also relevant to current events, holidays, or industry trends. For instance, ChatGPT can automatically create posts for important events, promotions during holidays, or timely announcements, ensuring that your audience stays engaged and informed.

The ability to inject timeliness into your social media strategy enhances your brand's agility and responsiveness. By incorporating relevant and current content, businesses can capture the attention of their audience and capitalize on trending topics. This dynamic approach to content creation allows your brand to stay in sync with the ever-changing social media landscape, contributing to increased engagement and interaction.

Multichannel Support

ChatGPT's versatility extends to supporting multiple social media platforms, enabling businesses to automate posts across various channels simultaneously. This multichannel support is instrumental in reaching a wider audience with a consistent message. Whether it's Facebook, Twitter, Instagram, or other platforms, ChatGPT can adapt to the unique requirements and characteristics of each, ensuring a seamless and cohesive presence.

Automating posts across different social media channels not only expands your brand's visibility but also maximizes the potential reach to diverse demographics. This broadened reach allows businesses to tap into new markets and connect with potential customers who may frequent specific platforms. The ability to maintain a consistent message across diverse channels reinforces brand cohesion and contributes to an overarching marketing strategy.

Increased Efficiency

Automating social media posts with ChatGPT translates to increased efficiency by streamlining the content creation process. By automating routine tasks, businesses can free up valuable time and resources that can be redirected toward more strategic aspects of their operations. This increased efficiency contributes to overall productivity and allows teams to focus on devising innovative marketing strategies, analyzing performance metrics, or engaging in direct customer interactions.

The efficiency gains realized through ChatGPT's automation capabilities are particularly beneficial for businesses of all sizes. Small businesses can operate with leaner teams, while larger corporations can optimize their social media workflows for maximum impact. The result is not only a reduction in operational bottlenecks but also an improvement in the quality and effectiveness of social media content.

In conclusion, automating social media posts with ChatGPT emerges as a transformative strategy for businesses seeking to enhance the consistency, relevance, and efficiency of their social media marketing efforts. The ability to maintain a consistent brand voice, generate relevant and timely content, achieve multichannel support, and increase overall efficiency positions ChatGPT as a versatile tool in optimizing social media communication.

Whether you're a small business looking to streamline your social media presence or a large corporation aiming to reach a diverse audience, ChatGPT offers a powerful solution. By leveraging automation, businesses can not only save time and resources but also elevate the quality of their social media content, leading to increased engagement, brand visibility, and ultimately, business success.

Answering Customer Inquiries

Responding to customer inquiries is a pivotal aspect of any business, and the effectiveness of this communication can significantly impact customer satisfaction and loyalty. ChatGPT offers a transformative solution by automating the process of addressing customer inquiries. This not only streamlines the response mechanism but also ensures quick, accurate, and personalized interactions, ultimately contributing to enhanced customer relationships and improved business outcomes.

Quick and Accurate Responses

ChatGPT's ability to generate quick and accurate responses is a game-changer for businesses aiming to provide timely support to their customers. In a fast-paced business environment, customers expect swift resolutions to their inquiries. ChatGPT excels in delivering rapid responses, allowing businesses to address customer concerns promptly.

The accuracy of responses is equally crucial for customer satisfaction. ChatGPT's proficiency in understanding the context and generating coherent, relevant replies ensures that customers receive precise information. Quick and accurate responses not only resolve issues efficiently but also contribute to an improved overall customer experience, building trust and loyalty.

Increased Efficiency

Automating customer inquiries with ChatGPT not only ensures quick responses but also significantly enhances efficiency by reducing the workload on customer service teams. By handling routine inquiries, ChatGPT allows human resources to be redirected toward more complex tasks, strategic planning, or addressing unique customer needs that require human intervention.

This increased efficiency is particularly beneficial for businesses dealing with a high volume of inquiries. Whether you are a small business with limited staff or a large corporation managing a vast customer base, ChatGPT's automation capabilities empower your team to optimize their efforts and deliver exceptional service. The result is improved productivity and a more streamlined customer support workflow.

Consistent Brand Voice

Maintaining a consistent brand voice across customer interactions is crucial for reinforcing brand identity and building trust. ChatGPT can be trained on your existing brand voice, ensuring that all automated responses align with your brand's tone, values, and language. This consistency contributes to a professional image and fosters trust with your customers.

The ability to replicate your brand voice in every customer interaction, whether automated or human-led, reinforces your brand's identity. Consistency in communication builds familiarity and reliability, instilling confidence in customers that their inquiries are handled with the same level of professionalism and care they have come to expect from your brand.

Multichannel Support

Customers today engage with businesses across various channels, including email, chat, and social media. ChatGPT's versatility allows it to respond to customer inquiries seamlessly across multiple channels. This multichannel support not only expands the reach of your customer service but also enhances accessibility for your customers.

Automating responses on different channels ensures that customers receive consistent and timely support, regardless of their preferred communication platform. This approach contributes to a holistic customer service strategy, catering to diverse audience preferences and maximizing your business's ability to connect with customers wherever they are.

Personalized Responses

Personalization is a key factor in delivering exceptional customer service. ChatGPT excels in generating personalized responses based on the specific nature of each customer's inquiry. By understanding the context and tailoring responses to individual needs, businesses can provide high-quality, individualized support.

Personalized responses contribute to building strong relationships with customers. Whether addressing product-related questions, providing assistance with issues, or offering personalized recommendations, ChatGPT ensures that customers feel valued and understood. This personalized touch goes a long way in enhancing customer satisfaction and fostering brand loyalty.

However, it is important to address the data privacy concerns related to personalization. Collecting and using customer data to create personalized experiences can raise significant privacy issues. To mitigate these concerns, businesses should implement privacy-by-design principles in the development and deployment of personalized marketing strategies using AI.

Privacy-by-Design Principles for Personalized Marketing

Data Minimization: Collect only the data that is necessary for personalization. Avoid gathering excessive or irrelevant information.

User Consent: Obtain explicit consent from users before collecting their data. Ensure that they understand what data is being collected and how it will be used.

Transparency: Be transparent about data collection and usage practices. Clearly communicate to users how their data is being used to personalize their experience.

Anonymization: Where possible, anonymize user data to protect their identity. This reduces the risk of privacy breaches.

Data Security: Implement robust security measures to protect user data from unauthorized access, breaches, or misuse. Regularly update these measures to address emerging threats.

User Control: Provide users with control over their data. Allow them to view, edit, or delete their data, and give them the option to opt out of data collection and personalized marketing.

Compliance: Ensure that your data practices comply with relevant data protection regulations, such as GDPR and CCPA.

Regular Audits: Conduct regular audits of your data collection and usage practices to ensure compliance with privacy standards and identify potential areas for improvement.

By incorporating these privacy-by-design principles, businesses can create personalized marketing strategies that not only enhance customer satisfaction but also respect and protect user privacy. This approach helps build trust with customers and fosters long-term loyalty while also safeguarding your brand's reputation.

Summary

In conclusion, the integration of ChatGPT into marketing tasks marks a transformative shift in the industry. By automating repetitive and mundane tasks, enhancing content creation, improving customer service, and facilitating data-driven personalization, ChatGPT empowers marketing professionals to elevate their strategies and achieve higher efficiency.

As explored throughout the chapter, ChatGPT proves invaluable in various facets of marketing, including content generation, customer service, market research, ad copywriting, email marketing, social media management, and lead generation. The technology's ability to understand context and produce coherent, human-like text ensures the creation of high-quality, relevant content that resonates with target audiences.

Moreover, ChatGPT's capabilities extend beyond text generation. It enhances visual content, supports localized marketing efforts, and provides strategic collaboration opportunities. By leveraging predictive analytics and AI-infused social listening, marketers can stay ahead of trends and make informed decisions.

Ethical considerations remain paramount in the integration of ChatGPT. Businesses must prioritize transparency, data security, and inclusivity while adhering to regulatory standards. By fostering a culture of continuous learning and staying attuned to evolving trends, businesses can position themselves as pioneers in the digital marketing landscape.

Ultimately, ChatGPT is not just a tool but a catalyst for innovation and transformation in marketing. As businesses embrace this AI-powered journey, they unlock the potential for improved outreach, engagement, and success. The strategic implementation of ChatGPT in daily operations will reshape marketing strategies, enabling professionals to navigate the complexities of their industry with agility and creativity.

CHAPTER 4

Advanced Techniques for Marketing with ChatGPT

As you delve deeper into your journey with ChatGPT, you'll find a plethora of advanced techniques waiting to be explored, each offering the potential to amplify your marketing endeavors to new heights. In this chapter, we'll explore some of the more important ones.

Reviewing ChatGPT's Advanced Techniques

With its versatility and adaptability, ChatGPT can seamlessly integrate into various marketing platforms and workflows, enhancing collaboration and streamlining processes across your marketing ecosystem. From customer relationship management (CRM) systems to social media management tools, ChatGPT can be seamlessly integrated, ensuring a cohesive and efficient marketing operation. This integration not only saves time and resources but also fosters a more agile and responsive marketing approach, enabling you to stay ahead of the curve in today's fast-paced digital landscape.

ChatGPT stands as a formidable ally on your marketing journey, offering a suite of advanced techniques designed to propel your efforts to new heights. Whether you're a seasoned marketer seeking to refine your strategies or a newcomer eager to make your mark, ChatGPT is poised to empower you every step of the way, unlocking a world of possibilities to help you achieve your marketing goals with finesse and efficiency.

© Eldar Najafov 2024
E. Najafov, *ChatGPT for Marketing*, https://doi.org/10.1007/979-8-8688-0312-3_4

Personalized Marketing

One of the most potent capabilities of ChatGPT lies in its ability to craft personalized marketing messages with precision. By harnessing customer data encompassing demographics, behaviors, and preferences, ChatGPT empowers you to tailor your marketing efforts on an individual level, significantly enhancing the relevance and impact of your campaigns. For example, ChatGPT can be used to send personalized emails based on past purchase behavior or to create targeted advertisements that speak directly to individual customer needs. Another example is using ChatGPT to tailor social media content that resonates with specific segments of your audience.

However, the use of personalized marketing also presents certain challenges. Chief among these are data privacy concerns and the potential overreliance on automated insights without human interpretation.

Data Privacy Concerns

When utilizing customer data for personalization, it is crucial to safeguard this information to maintain customer trust and comply with regulations such as GDPR and CCPA. Businesses must obtain explicit consent from customers before collecting and using their data for marketing purposes. Additionally, implementing robust data security measures to protect against breaches and unauthorized access is essential.

Best Practices for Safeguarding Customer Data

Obtain Explicit Consent: Always seek clear and explicit consent from customers before collecting and using their data. Provide transparent information about how their data will be used.

Data Anonymization: Where possible, anonymize customer data to minimize the risk of identifying individuals if a data breach occurs.

Regular Audits: Conduct regular audits of your data practices to ensure compliance with relevant regulations and to identify and mitigate any potential security risks.

Employee Training: Ensure that all employees handling customer data are trained in data privacy best practices and understand the importance of safeguarding customer information.

Blending AI Insights with Human Judgment

While AI, like ChatGPT, provides valuable insights and automates many aspects of personalized marketing, it is essential to balance this with human judgment. Automated insights should be reviewed and interpreted by human marketers to ensure they align with the overall strategy and brand values. This blend of AI and human oversight ensures that personalization efforts are both effective and ethical.

Ethical Personalization Practices

Human Oversight: Ensure that AI-generated insights and content are reviewed by human marketers to maintain brand consistency and ethical standards.

Transparency: Be transparent with customers about the use of AI in personalization efforts. Inform them how their data is being used and the benefits they can expect.

Bias Mitigation: Regularly review AI models to detect and mitigate any biases that may arise in the personalized content generated.

By following these best practices, businesses can harness the power of ChatGPT for personalized marketing while addressing data privacy concerns and ensuring ethical use of AI insights. This balanced approach not only enhances the effectiveness of marketing campaigns but also builds and maintains customer trust.

Data Analysis

ChatGPT isn't just a tool for communication—it's a powerhouse for data analysis. Armed with its natural language processing prowess, ChatGPT can sift through vast troves of customer data, distilling intricate patterns and insights that would otherwise remain elusive. This invaluable resource equips you with the intelligence needed to make informed, data-driven decisions that can drive your marketing strategies forward with confidence.

Content Generation

Gone are the days of wracking your brain for fresh content ideas or wrestling with writer's block. ChatGPT excels as a content generation powerhouse, effortlessly churning out a diverse array of marketing collateral—from engaging blog posts to compelling product descriptions—all with the swipe of a keyboard. This not only frees up valuable time and resources but also ensures a consistent stream of high-quality content to fuel your marketing endeavors.

Limits of AI in Capturing Brand Voice

While ChatGPT is adept at generating content, it may struggle to fully capture the unique voice and nuanced tone of a brand without human oversight. Relying solely on AI for content creation can result in generic or misaligned messages that do not resonate with the intended audience.

Case Studies

Company A: Ecommerce Blog Posts

Challenge: Company A needed to generate regular blog posts to drive traffic to their ecommerce site but struggled with consistent content creation.

Solution: By using ChatGPT, they automated the initial draft process of their blog posts.

Outcome: The marketing team reviewed and refined the AI-generated drafts, ensuring the content aligned with the brand's voice and messaging. This process significantly reduced the time spent on content creation while maintaining high quality.

Challenge Faced: Initially, the content produced lacked the brand's unique tone, requiring substantial edits.

Resolution: The team trained ChatGPT with examples of past content and established a review framework to integrate human oversight effectively.

Company B: Product Descriptions

Challenge: Company B needed to create compelling product descriptions for a large inventory but found it resource-intensive.

Solution: They deployed ChatGPT to generate the first drafts of product descriptions.

Outcome: The AI-produced descriptions provided a solid base, which the content team then customized to ensure each description reflected the brand's personality.

Challenge Faced: Some AI-generated descriptions were too generic and required human creativity to make them engaging.

Resolution: The team used a hybrid approach, combining AI efficiency with human creativity to achieve the desired quality.

Framework for Integrating Human Oversight

To ensure that AI-generated content aligns with brand values and captures the brand voice effectively, a structured framework for human oversight is essential.

Training the AI

Data Preparation: Provide ChatGPT with a robust dataset of past content that exemplifies the brand's voice and tone.

Guidelines: Establish clear guidelines on tone, style, and messaging for the AI to follow.

Initial Draft Creation

AI-Generated Drafts: Use ChatGPT to create initial drafts of content, leveraging its ability to produce consistent and high-quality text quickly.

Human Review and Refinement

Editorial Review: Human editors review AI-generated content to ensure it aligns with brand guidelines and resonates with the target audience.

Customization: Editors add nuanced elements and brand-specific touches that the AI might miss.

Feedback Loop

Iterative Training: Continuously train ChatGPT based on feedback from the editorial team to improve its understanding of the brand voice.

Performance Monitoring: Regularly assess the performance of AI-generated content and make adjustments to the training data and guidelines as needed.

Quality Assurance

Final Check: Implement a final quality assurance check to ensure that all content meets the highest standards before publication.

By following this framework, businesses can effectively integrate human oversight into the AI content generation process, ensuring that the content not only benefits from AI efficiency but also aligns with the brand's values and voice.

Lead Generation

Say goodbye to manual lead generation processes that consume precious hours of your time. With ChatGPT at the helm, automating lead generation becomes a seamless endeavor. Whether it's mining leads from social media platforms, orchestrating targeted email campaigns, or optimizing online forms, ChatGPT streamlines the process, enabling you to cast a wider net and connect with your audience more efficiently.

Potential Pitfalls of Overly Automated Interactions

While automation can significantly enhance efficiency, overly automated interactions may come across as impersonal to potential leads. It is essential to strike a balance between leveraging AI for efficiency and maintaining a personal touch to build genuine connections with your prospects.

Segmentation and Personalization

Segment Your Audience: Use ChatGPT to analyze data and segment your audience based on specific criteria such as demographics, behavior, and past interactions. This allows for more personalized communication.

Personalized Messages: Automate the initial outreach with personalized messages that address the recipient by name and reference specific needs or interests. Ensure the content is relevant and tailored to each segment.

Automated Follow-Ups with Human Oversight

Initial Follow-Up: Use ChatGPT to automate follow-up emails or messages after an initial contact, ensuring timely responses that keep the lead engaged.

Human Review: Have a human review the responses to identify leads that show high engagement or potential, allowing for more personalized and thoughtful follow-ups by a sales representative.

AI for Lead Scoring

Identify High-Potential Leads: Leverage ChatGPT to analyze interactions and behavior to score leads based on their likelihood to convert. This helps prioritize leads that should receive more personalized attention.

Human Interaction for High-Priority Leads: Use the lead scores to determine which leads should receive direct interaction from sales representatives, ensuring that high-potential leads get the personal touch needed to close the sale.

Blending AI and Human Interaction

AI-Assisted Conversations: Implement ChatGPT to assist with initial conversations, providing quick responses and gathering essential information. Ensure that these interactions feel natural and conversational.

Seamless Transition: Train ChatGPT to seamlessly transition the conversation to a human representative at the appropriate time, ensuring that the prospect feels valued and heard.

Scheduled Calls or Meetings: After initial automated interactions, schedule calls or meetings with a human representative to discuss the prospect's needs in detail.

Personalized Follow-Up Content: Send personalized follow-up content such as case studies, white papers, or product demos based on the prospect's specific interests and inquiries.

Techniques for Using AI and Human Interaction

AI for Data Collection: Use ChatGPT to gather and analyze data on prospects, providing valuable insights for the sales team.

Human Interaction for Relationship Building: Focus human efforts on building relationships and trust with prospects, addressing specific concerns, and providing personalized solutions.

Automated Nurturing Campaigns: Set up automated email nurturing campaigns with personalized content that guides prospects through the sales funnel, with human intervention at critical points to answer questions and provide support.

By implementing these strategies, businesses can maintain a balance between the efficiency of automation and the personal touch needed to build strong relationships with potential leads. This approach ensures that lead generation is both effective and human-centric, enhancing the overall customer experience and increasing conversion rates.

A/B Testing

The path to marketing success often involves navigating through a maze of strategies, each vying for the coveted title of the most effective approach. Enter ChatGPT, your trusted ally in the realm of A/B testing. By leveraging its analytical prowess, you can conduct rigorous experiments, comparing different marketing tactics with precision and granularity. Armed with actionable insights, you can fine-tune your strategies, optimize performance, and maximize ROI with unparalleled precision.

Combining AI-Driven A/B Testing with Human Insights

While AI-driven A/B testing offers significant advantages in terms of speed and efficiency, it is crucial to temper reliance on AI with human insights. AI systems, including ChatGPT, can sometimes miss the subtleties of customer behavior or interpret data with inherent biases. Here are strategies to effectively combine AI-driven A/B testing with human insights.

Initial Setup and Hypothesis Generation

Human-Led Hypothesis: Start with a human-led hypothesis generation process, where marketers use their understanding of the audience and market trends to formulate testable hypotheses.

AI-Driven Analysis: Use ChatGPT to analyze historical data and suggest potential variables and metrics for testing, providing a data-driven foundation for the hypotheses.

Running A/B Tests with AI

Automated Testing: Implement ChatGPT to automate the execution of A/B tests, ensuring that all variables are tested systematically and data is collected accurately.

Real-Time Analysis: Utilize ChatGPT's real-time analytical capabilities to monitor ongoing tests, providing instant insights into performance metrics and trends.

Interpreting Results

Human Interpretation: After the AI has processed the test data, have human marketers review the results. They can interpret nuances that the AI might overlook, such as context-specific factors or unexpected external influences.

Identifying Biases: Marketers should be vigilant in identifying any biases in the AI's analysis, ensuring that results are not skewed by data quality issues or algorithmic biases.

Continuous Training and Monitoring

Iterative Training: Continuously train ChatGPT with new data and feedback to improve its understanding and accuracy. This includes incorporating results from past A/B tests and adjusting the AI's algorithms to refine its predictive capabilities.

Monitoring for Accuracy: Regularly monitor the performance of ChatGPT to ensure it remains accurate and unbiased. Implement checks and balances to detect any deviations or errors in interpretation.

Understanding Customer Behavior

Qualitative Insights: Complement AI-driven quantitative analysis with qualitative insights gathered from customer feedback, surveys, and interviews. This provides a more holistic view of customer behavior and preferences.

Behavioral Analysis: Use human expertise to delve deeper into behavioral patterns that AI might not fully capture, such as emotional responses or cultural influences.

Blending AI Insights with Human Judgment

Strategic Decisions: Use AI-generated insights as a starting point for strategic decisions, but ensure that human judgment is applied to validate and contextualize these insights.

Adaptive Strategies: Be prepared to adapt strategies based on human observations and insights that may not be immediately evident in the AI's analysis.

Importance of Ongoing Training and Monitoring

The effectiveness of AI-driven A/B testing hinges on the continuous improvement of the AI system. This requires ongoing training and monitoring to ensure that ChatGPT remains accurate, unbiased, and aligned with your marketing objectives.

Regular Updates: Keep ChatGPT updated with the latest data and market trends to maintain its relevance and accuracy.

Bias Detection: Implement robust mechanisms to detect and correct biases in the AI's analysis, ensuring fair and unbiased results.

Human Oversight: Maintain a strong element of human oversight in all stages of A/B testing to ensure that AI-driven insights are appropriately validated and contextualized.

By effectively combining AI-driven A/B testing with human insights, businesses can leverage the strengths of both AI and human expertise. This balanced approach ensures that marketing strategies are not only data-driven but also nuanced and responsive to the complexities of customer behavior.

Integrating ChatGPT with Other Tools

Integrating ChatGPT with other tools can significantly enhance your marketing operations, enabling you to streamline workflows, automate mundane tasks, and harness valuable customer insights. Here's a deeper dive into how you can leverage ChatGPT's integration capabilities across various platforms.

CRM Integration

Seamlessly connecting ChatGPT with your customer relationship management (CRM) system empowers you with real-time access to customer data. By automating lead generation, personalized customer engagement, and targeted marketing campaigns, you can amplify customer satisfaction and bolster your marketing efficacy. Harnessing ChatGPT within your CRM framework not only optimizes customer interactions but also cultivates lasting relationships with your audience.

Specific Use Cases of ChatGPT Integration with CRM Systems

Automated Lead Generation and Scoring

Scenario: A mid-sized ecommerce company integrates ChatGPT with their CRM to automate lead generation.

Application: ChatGPT mines leads from social media interactions and website visits, automatically scoring them based on engagement metrics.

Outcome: The company saw a 30% increase in qualified leads and a 20% reduction in the time spent on manual lead generation tasks.

Personalized Customer Engagement

Scenario: A B2B service provider uses ChatGPT to enhance customer engagement through personalized communication.

Application: ChatGPT sends tailored emails and messages based on customer behavior and preferences stored in the CRM.

Outcome: This led to a 25% increase in email open rates and a 15% boost in customer retention rates.

Targeted Marketing Campaigns

Scenario: A retail chain integrates ChatGPT with their CRM to manage targeted marketing campaigns.

Application: ChatGPT analyzes customer purchase history and behavior, recommending personalized promotions and products.

Outcome: The retail chain experienced a 40% increase in campaign ROI and a 35% rise in customer engagement.

Implementation Guidance for Integrating ChatGPT with Popular CRM Systems

Step-by-Step Guide for SMEs

Select Your CRM System

Popular options include Salesforce, HubSpot, Zoho CRM, and Microsoft Dynamics 365. Choose a CRM that best suits your business needs and budget.

API Integration

Salesforce: Use the Salesforce API to connect ChatGPT. Follow Salesforce's API documentation to generate API keys and set up the integration.

HubSpot: Utilize HubSpot's API to link ChatGPT. Refer to HubSpot's developer resources for step-by-step instructions.

Zoho CRM: Leverage Zoho CRM's API for integration. Zoho provides detailed API guides to facilitate the process.

Microsoft Dynamics 365: Integrate using Microsoft Dynamics API. Microsoft's documentation offers comprehensive guidance on API setup.

Data Mapping

Map the data fields between ChatGPT and your CRM to ensure seamless data flow. Identify key data points such as customer contact details, engagement history, and sales pipeline stages.

Configuration

Configure ChatGPT to automate specific tasks within the CRM, such as lead generation, customer follow-ups, and campaign management. Use pre-built templates or customize workflows to match your business processes.

Testing

Conduct thorough testing to ensure the integration works smoothly. Test various scenarios, including lead generation, personalized messaging, and data synchronization.

Training

Train your team on using the integrated system. Provide resources and training sessions to help them understand how to leverage ChatGPT within the CRM effectively.

Monitoring and Optimization

Regularly monitor the integration to identify any issues or areas for improvement. Use analytics and feedback to optimize the system for better performance and efficiency.

By following these steps, SMEs can effectively integrate ChatGPT with their CRM systems, unlocking numerous benefits such as enhanced customer engagement, improved lead management, and more efficient marketing campaigns.

Social Media Integration

Integrating ChatGPT with social media platforms, including Facebook and Twitter, revolutionizes your social media management. Through automated posting schedules, real-time engagement monitoring, and personalized responses, ChatGPT empowers you to curate a dynamic online presence and forge meaningful connections with your audience. By leveraging ChatGPT's natural language processing capabilities, you can craft compelling content, drive user engagement, and cultivate brand advocacy across diverse social channels.

Breaking Down the Setup Process and Day-to-Day Management

Setup Process

Platform Selection

Choose the social media platforms most relevant to your audience. While Facebook and Twitter are popular globally, consider local preferences such as WeChat in China, VKontakte (VK) in Russia, or WhatsApp in many regions for more personalized engagement.

API Integration

Facebook: Use the Facebook Graph API to connect ChatGPT. Follow Facebook's developer documentation to create an app, generate access tokens, and configure the API. Twitter: Leverage the Twitter API for integration. Set up a developer account, create an app, and use the provided keys and tokens to connect ChatGPT. Local Platforms: For region-specific platforms like WeChat, refer to their official API documentation to understand the integration requirements.

Configuration

Configure ChatGPT to handle specific tasks such as posting schedules, responding to messages, and monitoring engagement. Use templates or custom settings to align with your social media strategy.

Content Mapping

Map the types of content ChatGPT will handle, such as posts, replies, and direct messages. Ensure that ChatGPT understands your brand voice and guidelines to maintain consistency.

Testing

Test the integration to ensure smooth operation. Conduct scenario-based testing to validate posting schedules, response accuracy, and engagement monitoring.

Day-to-Day Management

Automated Posting

Schedule posts in advance using ChatGPT's automated posting feature. This ensures a consistent posting cadence, allowing you to maintain an active presence without manual effort.

Real-Time Engagement

Monitor real-time interactions using ChatGPT's engagement tracking. ChatGPT can respond to comments, mentions, and direct messages instantly, keeping your audience engaged.

Personalized Responses

Configure ChatGPT to provide personalized responses based on user interactions. For example, if a user asks about product details, ChatGPT can provide tailored information or direct them to relevant resources.

Content Creation

Use ChatGPT to generate compelling content ideas, draft posts, and create engaging multimedia content. This helps keep your social media feed fresh and interesting.

Analytics and Reporting

Leverage ChatGPT's analytics capabilities to track the performance of your social media activities. Analyze metrics such as engagement rates, follower growth, and content performance to optimize your strategy.

Highlighting Local Relevance with Examples

Facebook in the United States and Europe

A fashion retailer uses ChatGPT to manage its Facebook page, automating posts about new collections, responding to customer inquiries in real time, and running targeted ad campaigns based on user behavior insights.

WeChat in China

A technology company integrates ChatGPT with WeChat to provide customer support, send promotional messages, and engage with users through interactive content like quizzes and polls, tailored to local preferences.

VK in Russia

An entertainment brand uses ChatGPT on VKontakte (VK) to share updates about events, interact with fans through comments and messages, and analyze user engagement to improve future content.

WhatsApp in Latin America

A healthcare provider utilizes ChatGPT on WhatsApp to send appointment reminders, answer health-related queries, and share wellness tips, ensuring timely and personalized communication with patients.

By breaking down the setup process and day-to-day management and highlighting examples from various social media platforms, businesses can better understand the practical applications and benefits of integrating ChatGPT with their social media operations. This approach ensures that the content is accessible and relevant to readers from diverse business backgrounds and regions.

Marketing Automation Tools

Pairing ChatGPT with marketing automation tools revolutionizes your campaign execution and customer engagement strategies. From orchestrating email campaigns and optimizing lead generation processes to delivering personalized content recommendations, ChatGPT seamlessly integrates with marketing automation platforms to streamline operations and amplify campaign performance. By automating routine marketing tasks and delivering hyper-personalized experiences, you can nurture leads, drive conversions, and propel business growth.

Step-by-Step Approach for Integrating ChatGPT with Marketing Platforms

Identify Your Needs

Determine what specific marketing tasks you want to automate with ChatGPT, such as email marketing, social media engagement, or lead generation. This helps in selecting the right tools and integration methods.

Choose a Marketing Automation Platform

Select a marketing automation platform that aligns with your business needs and budget. Popular options include Mailchimp, HubSpot, ActiveCampaign, and Marketo. Ensure the platform supports API integrations.

Set Up an Account

Create an account on your chosen marketing automation platform and familiarize yourself with its features. Most platforms offer tutorials and customer support to help you get started.

API Integration

Obtain API keys from your marketing automation platform. Use these keys to connect ChatGPT to the platform. Refer to the platform's API documentation for detailed instructions on generating and using API keys.

Mailchimp: Follow Mailchimp's API guide to create an app, generate an API key, and configure the integration.

HubSpot: Use HubSpot's API documentation to set up the integration, generate API tokens, and connect ChatGPT.

ActiveCampaign: Refer to ActiveCampaign's API resources to create an app, obtain API credentials, and integrate ChatGPT.

Marketo: Leverage Marketo's API documentation for integration steps, including app creation and API key generation.

Configure ChatGPT

Set up ChatGPT to perform specific tasks within the marketing platform. This includes defining workflows for email campaigns, lead scoring, content recommendations, and customer segmentation. Use templates or customize settings to match your marketing strategy.

Testing and Validation

Conduct thorough testing to ensure the integration works as expected. Test various scenarios, such as automated email sending, lead scoring accuracy, and content personalization. Validate that data flows correctly between ChatGPT and the marketing platform.

Launch and Monitor

Launch your automated marketing campaigns using ChatGPT. Continuously monitor performance metrics such as open rates, click-through rates, lead conversion rates, and engagement levels. Use these insights to optimize your strategies.

Scalable Solutions for Different Business Segments

Small Businesses and Startups

Budget-Friendly Platforms: Choose cost-effective marketing automation platforms like Mailchimp or ActiveCampaign, which offer free or affordable plans.

Simplified Workflows: Focus on automating basic tasks such as welcome emails, follow-up messages, and simple lead generation forms.

Gradual Expansion: Start with essential automation features and gradually expand as your business grows and your marketing needs evolve.

Mid-sized Businesses

Advanced Features: Utilize platforms like HubSpot or Marketo that offer more advanced features such as dynamic content, advanced segmentation, and multichannel campaigns.

Integration with CRM: Integrate ChatGPT with both your CRM and marketing automation platform to create a seamless data flow and enhance customer insights.

Customized Workflows: Develop customized workflows that align with your marketing strategy, focusing on personalized customer journeys and targeted campaigns.

Large Enterprises

Comprehensive Solutions: Opt for enterprise-grade platforms like Marketo or Salesforce Marketing Cloud that offer comprehensive solutions for large-scale marketing operations.

Complex Automation: Implement complex automation workflows that encompass multichannel campaigns, advanced analytics, and AI-driven insights.

Dedicated Teams: Allocate dedicated teams to manage and optimize your marketing automation and AI integration, ensuring continuous improvement and alignment with business goals.

By following these steps and considering scalable solutions tailored to your business size and needs, you can effectively integrate ChatGPT with marketing automation tools to enhance your campaign execution and customer engagement strategies. This approach ensures that businesses at various stages of their tech journey can leverage the power of AI to drive growth and success.

Analytics Tools

Integrating ChatGPT with robust analytics platforms empowers you to unlock actionable insights and drive data-informed marketing decisions. By harnessing ChatGPT's natural language processing capabilities, you can extract meaningful insights from vast datasets, track key performance metrics, and gauge the impact of your marketing initiatives with precision. From monitoring website traffic patterns and assessing conversion rates to evaluating customer sentiment and forecasting market trends, ChatGPT integration equips you with the analytical firepower needed to optimize your marketing strategies and drive sustainable growth.

Demystifying Analytics

Integrating ChatGPT with your analytics platforms can seem complex, but it's simpler than it appears. Here's how it works in basic terms.

ChatGPT can take the large amounts of data your business collects and help make sense of it. This means turning raw numbers into clear, understandable insights. For example, instead of manually sorting through thousands of customer feedback entries, ChatGPT can summarize the main points and trends for you.

Practical Examples with Visual Scenarios

Before Integration: Imagine you have a huge spreadsheet full of website traffic data. You spend hours looking at numbers, trying to figure out which pages are most popular and why some are not performing well.

After Integration: With ChatGPT, you can simply ask, "Which pages had the most traffic last month and why?" ChatGPT will analyze the data and provide you with an easy-to-understand report, highlighting the top pages and potential reasons for their performance, such as popular content or effective keywords.

Example Scenario

Website Traffic Analysis

Before: Manually reviewing thousands of rows in a spreadsheet.

After: ChatGPT provides a summary, "Your homepage and blog section are the most visited. High traffic on the blog is due to recent posts on trending topics."

Conversion Rates Assessment

Before: Struggling to correlate various data points to understand conversion rates.

After: ChatGPT analyzes and says, "Conversion rates increased by 15% last quarter, primarily due to the new email marketing campaign."

Customer Sentiment Evaluation

Before: Manually reading through customer reviews to gauge sentiment.

After: ChatGPT summarizes, "Customer sentiment is mostly positive this month, with frequent mentions of quick delivery and product quality."

How ChatGPT Enhances Data Analysis

Simplifies Complex Data: ChatGPT can break down complex datasets into easy-to-understand summaries. For example, if you have a large dataset from your social media platforms, ChatGPT can highlight key trends such as "Most customers are engaging with posts about sustainability."

Tracks Key Performance Metrics: ChatGPT helps you keep track of important metrics without getting lost in the data. You can ask it to monitor specific metrics like "monthly sales growth" or "weekly website visits," and it will provide regular updates.

Evaluates Marketing Impact: It can assess how well your marketing campaigns are performing. If you launch a new ad campaign, ChatGPT can analyze the data and tell you, "This campaign increased website traffic by 20%, particularly from mobile users."

By integrating ChatGPT with your analytics tools, you can make data analysis much more approachable and efficient. Whether you are new to data analytics or have some experience, ChatGPT can simplify the process, making it easier to understand and act on the data. This helps you make better marketing decisions, improve your strategies, and ultimately drive your business growth.

Content Management Systems (CMS) Integration

Seamlessly integrating ChatGPT with your CMS platform empowers you to automate content creation, optimize search engine optimization (SEO) strategies, and enhance user engagement. By leveraging ChatGPT's natural language generation capabilities, you can effortlessly generate high-quality blog posts, articles, and product descriptions tailored to your target audience's preferences. Furthermore, ChatGPT can assist in content ideation, headline optimization, and keyword research, enabling you to craft compelling narratives and drive organic traffic to your website. With ChatGPT integrated into your CMS workflow, you can streamline content production processes, improve content relevance, and deliver exceptional user experiences across all digital touchpoints.

Ecommerce Platforms Integration

Integrating ChatGPT with ecommerce platforms, such as Shopify, WooCommerce, or Magento, can supercharge your online retail operations. By deploying ChatGPT-powered chatbots, you can offer personalized product recommendations, provide real-time customer support, and facilitate seamless transactions, thereby enhancing the overall shopping experience. Moreover, ChatGPT can analyze customer feedback, predict purchasing trends, and optimize product listings to maximize conversion rates and drive revenue growth. With ChatGPT seamlessly integrated into your ecommerce ecosystem, you can cultivate customer loyalty, drive repeat purchases, and unlock new revenue streams with unparalleled efficiency.

Practical Implementation Steps

Platform Selection and API Integration

For Shopify, use Shopify's API to connect ChatGPT. Create a Shopify app, generate API credentials, and follow the integration guide.

For WooCommerce, utilize WooCommerce's REST API. Create API keys within the WooCommerce settings, then link ChatGPT to manage products and orders.

For Magento, leverage Magento's API. Set up an integration in the Magento admin panel, obtain API tokens, and configure the connection with ChatGPT.

Configuring ChatGPT for Specific Tasks

For personalized product recommendations, train ChatGPT to analyze customer browsing history and preferences. Configure it to suggest products based on this data, enhancing the shopping experience.

For real-time customer support, set up ChatGPT to handle common customer inquiries such as order status, return policies, and product information. This reduces the workload on human support agents and provides instant responses to customers.

For seamless transactions, integrate ChatGPT with payment gateways to facilitate smooth and secure transactions. Ensure that the chatbot can guide customers through the checkout process, addressing any issues that arise.

Testing and Launch

Conduct comprehensive testing to ensure all features work correctly. Test different scenarios like product searches, customer inquiries, and transaction flows to ensure a seamless user experience.

Once testing is complete, launch the ChatGPT integration and monitor its performance. Make adjustments as needed based on customer feedback and observed interactions.

Real-Life Examples and Case Studies

Personalized Product Recommendations

An online clothing retailer integrated ChatGPT with Shopify. The chatbot analyzed customer browsing history and previous purchases to suggest outfits and accessories. This personalized approach increased the average order value by 25% and boosted customer satisfaction.

Real-Time Customer Support

A tech gadgets store using WooCommerce integrated ChatGPT to handle customer inquiries. The chatbot provided instant answers to questions about product features, shipping times, and return policies. This reduced the number of support tickets by 40% and improved response times significantly.

Facilitating Seamless Transactions

A home decor store on Magento used ChatGPT to streamline the checkout process. The chatbot guided customers through each step, from selecting products to entering payment details, and addressed any issues in real time. This reduced cart abandonment rates by 30%.

Direct Benefits for End Users

Enhanced Shopping Experience: By offering personalized product recommendations, ChatGPT makes the shopping experience more relevant and enjoyable for customers. For example, if a customer frequently buys sports gear, the chatbot can suggest the latest sportswear and equipment tailored to their preferences.

Immediate Customer Support: Customers receive real-time assistance, which increases satisfaction and loyalty. If a customer needs help with an order, the chatbot can provide instant updates and solutions without the need for waiting.

Smooth and Secure Transactions: ChatGPT ensures that the transaction process is smooth and secure, addressing any issues that may arise and guiding customers through the process seamlessly. This leads to a more confident and hassle-free shopping experience.

By following these implementation steps and considering the provided examples, businesses can effectively integrate ChatGPT with their ecommerce platforms to enhance operations and improve customer experiences. This approach not only drives growth and efficiency but also ensures that customers receive personalized, timely, and secure interactions, ultimately fostering loyalty and repeat business.

Project Management Tools Integration

Leveraging ChatGPT's integration capabilities with project management tools like Asana, Trello, or Jira enables you to streamline collaboration, enhance productivity, and accelerate project delivery. By integrating ChatGPT-powered bots into your project management workflows, you can automate task assignments, schedule reminders, and facilitate communication among team members, fostering a culture of transparency and accountability. Additionally, ChatGPT can generate project reports, analyze workflow bottlenecks, and provide actionable insights to optimize resource allocation and project prioritization. With ChatGPT seamlessly integrated into your project management ecosystem, you can drive operational excellence, mitigate project risks, and achieve project milestones with precision.

Detailed Application Processes

Setup and Integration: To integrate ChatGPT with project management tools, begin by selecting the tool that best fits your team's needs. For Asana, create an API key in the developer console and link it with ChatGPT. For Trello, generate an API token from the Trello API settings, and for Jira, obtain the API credentials through the Jira admin panel. Use these credentials to configure ChatGPT and establish a secure connection.

Daily Operational Use: Once integrated, ChatGPT can perform a variety of tasks to streamline daily operations. It can automate task assignments by analyzing project requirements and team member availability. For example, when a new task is created in Asana, ChatGPT can automatically assign it to the most suitable team member based on their current workload and expertise. In Trello, ChatGPT can create and move cards across boards based on project progress and deadlines. In Jira, it can update issues, log work hours, and notify team members of changes in real time.

Automated Task Assignments: ChatGPT can analyze project data to automate task assignments, ensuring that tasks are allocated efficiently and according to team member strengths. This reduces the time spent on manual task allocation and helps prevent bottlenecks.

Scheduling Reminders: ChatGPT can schedule reminders for upcoming deadlines, meetings, and milestones. It can send notifications to team members via email or within the project management tool, helping to keep everyone on track and informed.

Facilitating Communication: By acting as a central communication hub, ChatGPT can facilitate conversations between team members. It can answer common questions, provide project updates, and relay important information, reducing the need for constant manual check-ins.

Quantifying Improvements

Productivity Improvements: Teams that have integrated ChatGPT into their project management workflows have reported significant improvements in productivity. For instance, a software development team using Jira observed a 20% reduction in time spent on administrative tasks, allowing team members to focus more on core project activities.

Project Delivery Times: Integrating ChatGPT has been shown to accelerate project delivery times. A marketing agency using Asana reported a 15% decrease in project turnaround time due to more efficient task assignments and real-time progress tracking.

Enhanced Collaboration: Teams using Trello with ChatGPT noted a 25% improvement in collaborative efficiency. The automated updates and reminders ensured that team members were always aligned and aware of project statuses, reducing delays caused by miscommunication.

Practical Examples

Task Assignments in Asana: An IT company integrated ChatGPT with Asana to manage their software development projects. ChatGPT analyzed task requirements and team member skill sets to automatically assign tasks. This led to a 30% increase in task completion rates and a 10% reduction in project delays.

Workflow Optimization in Trello: A digital marketing firm used ChatGPT with Trello to optimize their content creation process. ChatGPT automated the movement of cards based on content approval stages, resulting in a 20% increase in content production efficiency.

Issue Tracking in Jira: A product development team integrated ChatGPT with Jira to handle issue tracking and reporting. ChatGPT provided real-time updates and automatically logged work hours, which improved overall project tracking accuracy by 15%.

Integrating ChatGPT with project management tools like Asana, Trello, and Jira can significantly enhance productivity, streamline workflows, and improve project delivery times. By detailing the application processes and quantifying the benefits, businesses can better understand how to implement and leverage ChatGPT to achieve operational excellence and project success. Whether automating task assignments, scheduling reminders, or facilitating communication, ChatGPT offers a powerful solution to optimize project management and drive business growth.

Using ChatGPT for Data Analysis and Reporting

As businesses navigate the ever-expanding landscape of data, the quest for efficient and innovative solutions to extract actionable insights has become paramount. In this era of information overload, traditional methods of data analysis often fall short in handling the complexity and scale of modern datasets. Enter ChatGPT—a cutting-edge technology poised to revolutionize the field of data analysis and reporting.

ChatGPT's prowess in processing and comprehending vast amounts of data stems from its advanced algorithms and natural language processing capabilities. Unlike conventional tools, which may struggle with unstructured or nuanced data, ChatGPT excels in deciphering complex patterns and extracting meaningful information. Whether it's parsing through customer feedback, analyzing market trends, or monitoring social media sentiment, ChatGPT empowers businesses to gain valuable insights with unparalleled speed and accuracy.

The integration of ChatGPT into data analysis workflows offers a myriad of benefits, chief among them being the ability to make data-driven decisions with confidence. By leveraging ChatGPT's analytical capabilities, businesses can uncover hidden correlations, identify emerging trends, and anticipate market shifts in real time. This proactive approach not only enhances strategic planning but also enables organizations to seize opportunities and mitigate risks more effectively.

Simplifying Terminology

Hidden Correlations: These are connections or relationships between different pieces of data that aren't immediately obvious, for example, discovering that increased social media activity is linked to higher sales.

Market Shifts: Changes in the market environment, such as new trends, customer preferences, or economic conditions. Being able to anticipate these shifts helps businesses stay ahead of competitors.

How ChatGPT Enhances Data Analysis

Simplifies Complex Data: ChatGPT can take large amounts of data and break it down into easy-to-understand summaries. For instance, if you have data on customer purchases, ChatGPT can highlight which products are most popular and why.

Tracks Key Performance Metrics: ChatGPT helps keep track of important business metrics like monthly sales growth or website visits. It provides regular updates and insights, making it easier to understand how your business is performing.

Evaluates Marketing Impact: ChatGPT can assess how well your marketing campaigns are doing. For example, it can analyze data from a recent ad campaign and tell you how much it increased website traffic or sales.

Visualizing the Process with ChatGPT

Here's a simple flowchart to illustrate how ChatGPT fits into a data analysis workflow:

Data Collection ➤ Data Processing ➤ ChatGPT Analysis ➤ Actionable Insights ➤ Strategic Decisions

Moreover, ChatGPT's ability to automate routine data analysis tasks significantly streamlines the process, allowing teams to focus their time and resources on higher-value activities. Tasks such as data cleansing, where inaccuracies and inconsistencies are identified and rectified, are performed seamlessly by ChatGPT, reducing the likelihood of errors and ensuring data integrity. Similarly, data aggregation—the process of consolidating information from disparate sources—is expedited, enabling faster decision-making and more comprehensive insights.

Real-World Applications

Data Cleansing: Imagine a retail company with a large customer database. Over time, this database may accumulate errors such as duplicate entries, incorrect customer information, or outdated records. Manually identifying and correcting these errors can be time-consuming and prone to mistakes. ChatGPT can automate this process by scanning the database for inconsistencies, correcting errors, and removing duplicates. This ensures that the data is accurate and up-to-date, which is crucial for effective marketing and customer relationship management.

Example Scenario: A marketing team needs to clean a customer list before launching a new email campaign. ChatGPT analyzes the list, identifies and merges duplicate entries, corrects formatting errors, and updates outdated contact information. This not only saves hours of manual work but also improves the accuracy of the campaign, leading to higher engagement rates.

Data Aggregation: Consider a financial services company that collects data from multiple sources, including customer transactions, market trends, and economic indicators. Aggregating this data manually can be complex and error-prone. ChatGPT can automate the aggregation process by consolidating data from various sources into a single, coherent dataset. This enables analysts to quickly access comprehensive insights and make informed decisions.

Example Scenario: A financial analyst needs to prepare a quarterly report on market trends. ChatGPT gathers data from different financial databases, normalizes the information, and compiles it into a comprehensive report. This speeds up the reporting process and ensures that the data is consistent and reliable.

Focus on Business Impact

Time Savings: Automating routine tasks like data cleansing and aggregation can save significant amounts of time. For instance, a team that previously spent days cleaning and organizing data can now complete these tasks in a matter of hours with ChatGPT. This frees up valuable time for team members to focus on strategic activities such as data analysis and decision-making.

Example Scenario: A retail company that integrates ChatGPT for data cleansing reduces the time spent on this task by 80%, allowing the team to focus on analyzing customer behavior and developing targeted marketing strategies.

Error Reduction: Manual data handling is often prone to human errors, which can lead to inaccurate insights and poor business decisions. ChatGPT minimizes these errors by consistently applying data rules and standards. This ensures higher data integrity and reliability.

Example Scenario: A healthcare provider uses ChatGPT to aggregate patient data from various clinics. By automating this process, the provider ensures that patient records are accurate and complete, reducing the risk of medical errors and improving patient care.

Faster Time-to-Market: By streamlining data processing tasks, ChatGPT enables faster decision-making and quicker implementation of data-driven initiatives. This can lead to a shorter time-to-market for new products or services, giving businesses a competitive edge.

Example Scenario: A tech startup uses ChatGPT to process user feedback and market data, allowing them to quickly iterate and release new features. This accelerates their development cycle and helps them stay ahead of competitors.

By automating routine data analysis tasks, ChatGPT not only saves time and reduces errors but also enhances the overall efficiency of data management processes. This allows businesses to focus on higher-value activities and make faster, more informed decisions. The real-world applications and business impacts of ChatGPT's automation capabilities demonstrate its potential to transform data management and drive significant business outcomes.

Through its natural language processing capabilities, ChatGPT transforms raw data into cohesive narratives, weaving together key findings, trends, and recommendations into insightful reports. Whether it's generating executive summaries, trend analyses, or predictive insights, ChatGPT ensures that stakeholders receive timely and relevant information to support decision-making processes. This not only enhances transparency and accountability within organizations but also fosters a culture of data-driven decision-making at all levels.

In conclusion, the integration of ChatGPT into data analysis and reporting workflows represents a paradigm shift in the way businesses harness the power of data. Its advanced algorithms, coupled with its natural language processing capabilities, empower organizations to unlock actionable insights from complex datasets with unprecedented speed and accuracy. By automating routine tasks and streamlining the reporting process, ChatGPT enables businesses to make data-driven decisions with confidence, driving innovation and growth in an increasingly competitive landscape.

Summary

Integrating ChatGPT with various marketing tools and platforms offers significant benefits across different areas of your business operations. By seamlessly connecting ChatGPT with your CRM system, marketing automation tools, social media platforms, ecommerce platforms, and project management tools, you can enhance productivity, streamline processes, and improve customer engagement.

CRM Integration: Connecting ChatGPT with your CRM system provides real-time access to customer data, automating lead generation, personalized customer engagement, and targeted marketing campaigns. This optimizes customer interactions and cultivates lasting relationships.

Marketing Automation Tools: Pairing ChatGPT with marketing automation platforms revolutionizes campaign execution and customer engagement strategies. ChatGPT automates email campaigns, optimizes lead generation processes, and delivers personalized content recommendations, significantly enhancing campaign performance.

Social Media Integration: Integrating ChatGPT with social media platforms like Facebook and Twitter revolutionizes social media management. ChatGPT automates posting schedules, monitors real-time engagement, and provides personalized responses, enhancing your online presence and audience connection.

Ecommerce Platforms: Integrating ChatGPT with ecommerce platforms such as Shopify, WooCommerce, or Magento can supercharge online retail operations. ChatGPT-powered chatbots offer personalized product recommendations and real-time customer support and facilitate seamless transactions, enhancing the overall shopping experience and driving revenue growth.

Project Management Tools: Leveraging ChatGPT with project management tools like Asana, Trello, or Jira streamlines collaboration, enhances productivity, and accelerates project delivery. ChatGPT automates task assignments, schedules reminders, and facilitates communication among team members, fostering transparency and accountability.

Data Analysis: ChatGPT automates routine data analysis tasks such as data cleansing and aggregation, reducing errors and ensuring data integrity. This enables faster decision-making and more comprehensive insights, allowing teams to focus on higher-value activities.

Practical Benefits and Real-World Applications

By automating routine tasks, ChatGPT saves significant amounts of time and reduces the likelihood of human error. For example, in data management, ChatGPT can streamline data cleansing and aggregation, leading to faster time-to-market for data-driven products or services. In project management, automating task assignments and scheduling reminders with ChatGPT can result in a 20% improvement in productivity.

Real-world applications highlight the tangible benefits of integrating ChatGPT:

Personalized Product Recommendations: An online retailer saw a 25% increase in average order value by using ChatGPT to analyze customer behavior and suggest products.

Real-Time Customer Support: A tech store reduced support tickets by 40% with ChatGPT handling common inquiries.

Facilitating Seamless Transactions: A home decor store decreased cart abandonment rates by 30% by using ChatGPT to guide customers through the checkout process.

Integrating ChatGPT across various business functions can drive significant improvements in efficiency, customer engagement, and strategic decision-making. By leveraging ChatGPT's capabilities, businesses can streamline operations, reduce errors, and focus on high-value activities, ultimately driving growth and success.

CHAPTER 5

Insights Unveiled: ChatGPT and the Art of Marketing Data Analysis

In the rapidly evolving landscape of digital marketing, data is the new gold. Every interaction, every click, every transaction generates a wealth of information that, if properly analyzed, can provide invaluable insights into consumer behavior and market trends. However, the sheer volume of data can be overwhelming, making it challenging to extract meaningful insights. This is where AI, particularly ChatGPT, comes into play. This chapter explores how ChatGPT, an advanced AI model, is revolutionizing the way we analyze marketing data, providing businesses with actionable summaries that drive strategy and growth.

The digital age has transformed the way businesses operate. With every interaction leaving a digital footprint, the potential to understand customer behavior and market trends has never been greater. The key lies in effectively analyzing this data to derive meaningful insights. Traditional data analysis methods, while useful, often fall short in handling the sheer volume and complexity of data generated today. Enter ChatGPT—an advanced AI model designed to understand and generate human-like text, making it a powerful tool for analyzing and summarizing complex datasets.

ChatGPT's ability to process vast amounts of data and provide actionable insights is revolutionizing marketing strategies. It allows businesses to understand their customers better, optimize their campaigns, and stay ahead of market trends. This chapter delves into the specifics of how ChatGPT is used in marketing data analysis, the process involved, real-world applications, challenges, and future prospects.

© Eldar Najafov 2024
E. Najafov, *ChatGPT for Marketing*, https://doi.org/10.1007/979-8-8688-0312-3_5

The Role of Data in Modern Marketing

Marketing today is driven by data. From customer demographics and behavior patterns to campaign performance metrics, data informs every aspect of marketing strategies. Understanding this data allows businesses to tailor their products, services, and marketing efforts to meet the needs and preferences of their target audiences. However, the abundance of data available can be a double-edged sword. While it provides a wealth of information, it also requires sophisticated tools and techniques to analyze and interpret it effectively.

The vast array of data sources available today includes social media interactions, website analytics, email campaign metrics, customer reviews, and more. Each of these sources offers unique insights into different aspects of consumer behavior. For instance, social media data can reveal what customers are saying about a brand in real time, while website analytics can show how customers are interacting with a company's online presence. Email campaign metrics can provide insights into what types of content resonate with customers and drive engagement.

Effective data analysis can help businesses identify trends, measure the effectiveness of marketing campaigns, understand customer preferences, and make informed decisions. However, this requires advanced tools and techniques that can handle large datasets and provide meaningful insights. Traditional data analysis methods often involve manual processes that are time-consuming and prone to errors. Moreover, they may not be able to capture the nuances of unstructured data, such as customer reviews and social media posts.

This is where AI and machine learning come into play. These technologies can automate the data analysis process, handle large datasets, and provide more accurate and actionable insights. ChatGPT, with its advanced natural language processing capabilities, can analyze unstructured data and provide deeper insights into consumer behavior and market trends.

ChatGPT: A Brief Overview

ChatGPT, developed by OpenAI, is a language model that can understand and generate human-like text based on the input it receives. Its capabilities extend beyond simple text generation; it can analyze and summarize complex datasets, making it an invaluable tool for marketers. By leveraging natural language processing (NLP) and machine learning algorithms, ChatGPT can process vast amounts of data and provide insights that would otherwise require significant time and expertise to uncover.

At its core, ChatGPT is built on the principles of deep learning—a subset of machine learning that uses neural networks with many layers (hence "deep") to model complex patterns in data. This allows ChatGPT to understand the context and nuances of the input text, generate relevant responses, and even summarize large volumes of information.

ChatGPT's ability to understand and generate human-like text is powered by a massive dataset it was trained on. This dataset includes a diverse range of texts from books, articles, websites, and more. During training, the model learns to predict the next word in a sentence, given the preceding words. Over time, this allows it to generate coherent and contextually relevant text.

One of the key strengths of ChatGPT is its versatility. It can be used for a wide range of applications, from generating content and answering questions to analyzing and summarizing data. In the context of marketing data analysis, ChatGPT can process and interpret large volumes of data, identify trends and patterns, and generate actionable insights.

Data Collection and Preprocessing

The first step in using ChatGPT for marketing data analysis involves gathering and preprocessing the data. This includes cleaning the data, handling missing values, and normalizing it for analysis. Data preprocessing is crucial because it ensures that the data is in a suitable format for analysis and that any noise or errors are minimized. This step often involves

> **Data Cleaning**: Removing duplicates, correcting errors, and ensuring consistency in the dataset
>
> **Data Transformation**: Converting data into a usable format, such as normalizing values or encoding categorical variables
>
> **Data Integration**: Combining data from multiple sources to provide a comprehensive view of the information available

Effective data collection and preprocessing are critical for accurate analysis. Data cleaning involves identifying and correcting errors or inconsistencies in the data. This could include removing duplicate entries, correcting typos, or standardizing formats. For instance, dates might be stored in different formats across different datasets, and standardizing these formats is essential for accurate analysis.

Data transformation involves converting data into a format that can be easily analyzed. This could include normalizing values to ensure they fall within a specific range or encoding categorical variables into numerical values. For example, if a dataset includes a column for "customer satisfaction" with values such as "very satisfied," "satisfied," and "unsatisfied," these categorical values can be encoded as numerical values for analysis.

Data integration involves combining data from multiple sources to provide a comprehensive view of the information available. This could include merging data from different databases, combining online and offline data, or integrating data from various marketing channels. For instance, combining website analytics with social media data can provide a more holistic view of customer behavior.

Natural Language Processing (NLP)

ChatGPT leverages NLP techniques to understand the context and nuances of the data. It can interpret customer feedback, social media interactions, and other textual data to identify trends and patterns. For instance, it can analyze sentiment in customer reviews, identify common themes in social media posts, and extract key phrases from survey responses. By understanding the language and context of the data, ChatGPT can provide deeper insights into consumer behavior and preferences.

NLP is a field of AI that focuses on the interaction between computers and human language. It involves the development of algorithms that can understand, interpret, and generate human language. In the context of marketing data analysis, NLP can be used to analyze unstructured data, such as customer reviews, social media posts, and survey responses.

Sentiment analysis is one of the key applications of NLP in marketing data analysis. It involves determining the sentiment expressed in a piece of text, such as whether a customer review is positive, negative, or neutral. By analyzing sentiment in customer feedback, businesses can gain insights into how customers feel about their products and services.

Theme identification is another important application of NLP. It involves identifying common themes or topics in a collection of texts. For instance, by analyzing social media posts, businesses can identify what customers are talking about and what issues or topics are trending.

Key phrase extraction involves identifying the most important phrases or terms in a piece of text. This can help businesses understand what aspects of their products or services are most important to customers.

By leveraging NLP techniques, ChatGPT can analyze large volumes of unstructured data and provide deeper insights into consumer behavior and preferences. This can help businesses identify trends, understand customer sentiment, and make informed decisions.

Summarization and Insights Generation

After analyzing the data, ChatGPT generates summaries that highlight key insights. These summaries can cover various aspects, such as customer sentiment, campaign performance, and market trends. For example, it might identify that a particular product is receiving positive feedback for its quality but negative feedback for its price. It can also highlight emerging trends, such as increasing interest in sustainable products or a shift in consumer preferences toward online shopping. By providing concise and actionable insights, ChatGPT enables marketers to make informed decisions quickly and efficiently.

Summarization is a powerful feature of ChatGPT. It involves condensing large volumes of information into concise summaries that highlight the most important insights. This can save businesses significant time and effort in analyzing data and identifying key trends.

For instance, ChatGPT can analyze customer reviews and generate a summary that highlights the main points of feedback. This could include identifying common themes, such as customers praising the quality of a product but expressing concerns about its price. This can help businesses understand what aspects of their products are most valued by customers and where there may be opportunities for improvement.

ChatGPT can also analyze the performance of marketing campaigns and generate summaries that highlight key metrics, such as conversion rates, click-through rates, and ROI. This can help businesses understand which campaigns are most effective and where there may be opportunities to optimize their marketing efforts.

In addition to summarizing data, ChatGPT can generate actionable insights that help businesses make informed decisions. For instance, by analyzing social media data, ChatGPT can identify emerging trends, such as increasing interest in sustainable products or a shift in consumer preferences toward online shopping. These insights can help businesses stay ahead of market trends and make data-driven decisions.

Visualization and Reporting

The final step involves presenting the insights in a user-friendly format. ChatGPT can integrate with visualization tools to create comprehensive reports and dashboards, making it easier for marketers to interpret the findings. Visualization is a powerful tool because it allows users to see patterns and trends at a glance. For instance, a dashboard might display key performance indicators (KPIs) such as conversion rates, customer acquisition costs, and return on investment (ROI). By providing clear and visually appealing reports, ChatGPT helps marketers communicate their findings to stakeholders and make data-driven decisions.

Visualization is an essential aspect of data analysis. It involves presenting data in a visual format, such as charts, graphs, and dashboards, to make it easier to understand and interpret. Effective visualization can help businesses identify patterns and trends that may not be immediately apparent in raw data.

For instance, a dashboard that displays key performance indicators (KPIs) such as conversion rates, customer acquisition costs, and ROI can provide a quick overview of the performance of marketing campaigns. This can help businesses understand which campaigns are most effective and where there may be opportunities to optimize their marketing efforts.

Visualization tools can also help businesses communicate their findings to stakeholders. For instance, a visually appealing report that highlights key insights and trends can help businesses effectively communicate their findings to decision-makers and stakeholders. This can help ensure that data-driven insights are used to inform business decisions and drive strategy.

ChatGPT can integrate with various visualization tools to create comprehensive reports and dashboards. This can help businesses effectively present their findings and make data-driven decisions. By providing clear and visually appealing reports, ChatGPT helps businesses communicate their findings to stakeholders and make data-driven decisions.

Case Studies: Real-World Applications

Several companies have already started leveraging ChatGPT for marketing data analysis. This section delves into a few case studies, showcasing how businesses have benefited from this technology.

Case Study 1: Ecommerce Giant

An ecommerce giant implemented ChatGPT to analyze customer reviews and social media mentions. By processing this unstructured data, the company identified key areas for improvement in their product offerings and customer service. For instance, ChatGPT highlighted that many customers were dissatisfied with the delivery times. Armed with this insight, the company optimized its logistics operations, resulting in a significant increase in customer satisfaction and repeat purchases.

In this case, the ecommerce giant was able to leverage ChatGPT's NLP capabilities to analyze large volumes of unstructured data, such as customer reviews and social media mentions. By identifying key areas for improvement, the company was able to make data-driven decisions that improved customer satisfaction and drove business growth.

Case Study 2: Financial Services Firm

A financial services firm used ChatGPT to analyze the performance of its marketing campaigns. By examining data from various channels, including email, social media, and search engine marketing, ChatGPT provided a detailed analysis of which campaigns were most effective. The firm was able to relocate its marketing budget to focus on the most successful strategies, resulting in a higher ROI and increased customer acquisition.

In this case, the financial services firm was able to leverage ChatGPT's data analysis capabilities to gain insights into the performance of its marketing campaigns. By identifying the most effective strategies, the firm was able to optimize its marketing efforts and achieve a higher ROI.

Case Study 3: Consumer Goods Manufacturer

A consumer goods manufacturer utilized ChatGPT to understand market trends and consumer preferences. By analyzing sales data, customer feedback, and competitor information, ChatGPT identified a growing demand for eco-friendly products. The

113

manufacturer responded by launching a new line of sustainable products, which quickly became a bestseller and enhanced the company's brand image.

In this case, the consumer goods manufacturer was able to leverage ChatGPT's ability to analyze large volumes of data to identify emerging market trends. By responding to the growing demand for eco-friendly products, the company was able to capitalize on this trend and drive business growth.

Challenges and Considerations

While ChatGPT offers significant advantages, it's essential to address potential challenges, such as data privacy concerns, biases in AI, and the need for continuous model updates to ensure accuracy.

Data Privacy and Security

One of the primary concerns with using AI for data analysis is ensuring the privacy and security of the data. Businesses must comply with regulations such as the General Data Protection Regulation (GDPR) and the California Consumer Privacy Act (CCPA) to protect customer data. This involves implementing robust data encryption, anonymization techniques, and access controls to prevent unauthorized access and data breaches.

Data privacy and security are critical considerations when using AI for data analysis. Businesses must ensure that they comply with relevant regulations and implement robust security measures to protect customer data. This includes data encryption, anonymization techniques, and access controls to prevent unauthorized access and data breaches.

Bias and Fairness in AI

AI models, including ChatGPT, can sometimes exhibit biases based on the data they are trained on. For instance, if the training data contains biased or unrepresentative samples, the AI model may produce biased results. To mitigate this, it's crucial to use diverse and representative datasets for training and to regularly audit the model's outputs for fairness and accuracy.

Bias and fairness are important considerations when using AI for data analysis. AI models can sometimes exhibit biases based on the data they are trained on. This can lead to biased results that may not accurately represent the true nature of the data. To mitigate this, it's crucial to use diverse and representative datasets for training and to regularly audit the model's outputs for fairness and accuracy.

Continuous Model Updates

The digital marketing landscape is constantly evolving, and so too must the AI models used for data analysis. Continuous updates and retraining are necessary to ensure that ChatGPT remains accurate and relevant. This involves regularly incorporating new data, refining algorithms, and testing the model's performance to maintain its effectiveness.

Continuous model updates are essential to ensure that AI models remain accurate and relevant. The digital marketing landscape is constantly evolving, and AI models must be regularly updated to reflect these changes. This involves incorporating new data, refining algorithms, and testing the model's performance to maintain its effectiveness.

Future Prospects

The future of marketing data analysis looks promising with advancements in AI. ChatGPT and similar models will continue to evolve, offering even more sophisticated tools for marketers to gain a competitive edge.

Enhanced Personalization

As AI technology advances, the ability to provide highly personalized marketing experiences will improve. ChatGPT can analyze individual customer data to create tailored marketing messages, product recommendations, and offers. This level of personalization can significantly enhance customer engagement and loyalty.

Enhanced personalization is one of the key benefits of AI in marketing. As AI technology advances, the ability to provide highly personalized marketing experiences will improve. ChatGPT can analyze individual customer data to create tailored marketing messages, product recommendations, and offers. This level of personalization can significantly enhance customer engagement and loyalty.

Predictive Analytics

Future iterations of ChatGPT will likely incorporate more advanced predictive analytics capabilities. By analyzing historical data and identifying patterns, AI can predict future trends and consumer behaviors. This enables marketers to proactively adjust their strategies and stay ahead of the competition.

Predictive analytics is another key area where AI can provide significant benefits.

Integration with Other AI Technologies

The integration of ChatGPT with other AI technologies, such as computer vision and machine learning, will create even more powerful marketing tools. For example, combining NLP with image recognition can enhance the analysis of visual content, such as social media images and videos. This multimodal approach will provide a more comprehensive understanding of consumer preferences and behaviors.

Integration with other AI technologies is another exciting prospect for the future of marketing data analysis. By combining ChatGPT with other AI technologies, such as computer vision and machine learning, businesses can create even more powerful marketing tools. For example, combining NLP with image recognition can enhance the analysis of visual content, such as social media images and videos. This multimodal approach will provide a more comprehensive understanding of consumer preferences and behaviors.

Real-Time Analysis

Advancements in AI will also enable real-time data analysis, allowing marketers to respond to emerging trends and opportunities instantly. Real-time insights can drive dynamic marketing strategies, such as adjusting ad placements based on current events or optimizing pricing based on real-time demand.

Real-time analysis is another key area where AI can provide significant benefits. Advancements in AI will enable real-time data analysis, allowing marketers to respond to emerging trends and opportunities instantly. Real-time insights can drive dynamic marketing strategies, such as adjusting ad placements based on current events or optimizing pricing based on real-time demand.

Ethical AI and Bias Mitigation

As AI becomes more integral to marketing strategies, addressing ethical concerns and mitigating biases will be crucial. Future developments in ChatGPT and similar models will likely include more robust frameworks for ensuring fairness, transparency, and accountability in AI-driven decisions. This will involve continuous monitoring and updating of models to eliminate biases and ensure equitable treatment of all customer segments.

Greater Accessibility and Usability

AI tools will become more accessible and user-friendly, enabling marketers with varying levels of technical expertise to leverage these technologies. This democratization of AI will empower small and medium-sized businesses to compete with larger enterprises by utilizing advanced data analysis tools. User interfaces will be designed to be intuitive, with easy integration into existing marketing platforms.

Cross-Channel Integration

Future AI developments will facilitate better integration across multiple marketing channels, providing a unified view of customer interactions. This will enable marketers to deliver consistent and cohesive experiences across various touchpoints, from social media and email campaigns to in-store interactions. AI-driven insights will help in orchestrating multichannel campaigns that are more aligned with customer journeys.

Sustainability and Social Responsibility

With growing emphasis on sustainability and social responsibility, AI will play a critical role in helping businesses align their marketing strategies with these values. ChatGPT can analyze consumer sentiment regarding sustainability issues and help brands develop more responsible marketing messages. Predictive analytics can also assist in forecasting the impact of sustainable practices on brand loyalty and sales.

In conclusion, the integration of ChatGPT into marketing data analysis processes can transform how businesses approach their strategies. By providing detailed, actionable insights, this AI technology enables marketers to make data-driven decisions with confidence. As AI continues to advance, the potential applications and benefits will only grow, making ChatGPT an indispensable tool for modern marketers.

By harnessing the power of AI, businesses can not only keep pace with the rapidly changing digital landscape but also lead the way in innovation and creativity. The future holds exciting possibilities, and the journey has just begun.

Summary

Incorporating ChatGPT into marketing data analysis processes can revolutionize how businesses approach their strategies. By providing detailed, actionable insights, this AI technology enables marketers to make data-driven decisions with confidence. As AI continues to advance, the potential applications and benefits will only grow, making ChatGPT an indispensable tool for modern marketers.

Detailed Summary

1. **Transforming Marketing Strategies**: ChatGPT's ability to process vast amounts of data and generate actionable insights is reshaping marketing strategies. Businesses can gain a deeper understanding of customer behavior, optimize their campaigns, and stay ahead of market trends. This transformation is essential in the current digital age, where data-driven decisions are crucial for success.

2. **Comprehensive Data Analysis**: The chapter explored the various steps involved in using ChatGPT for marketing data analysis:

   - **Data Collection and Preprocessing**: Gathering and preparing data to ensure accuracy and usability

   - **Natural Language Processing (NLP)**: Analyzing unstructured data such as customer reviews and social media interactions to uncover trends and sentiments

- **Summarization and Insights Generation**: Condensing large volumes of information into concise, actionable summaries that highlight key insights

- **Visualization and Reporting**: Presenting insights in a user-friendly format, making it easier for marketers to interpret and act on the findings

3. **Real-World Applications**: Several case studies demonstrated the real-world impact of ChatGPT on businesses:

   - An ecommerce giant improved customer satisfaction by optimizing logistics operations based on insights from customer reviews and social media mentions.

   - A financial services firm achieved a higher ROI by reallocating marketing budgets to the most effective strategies.

   - A consumer goods manufacturer capitalized on the growing demand for eco-friendly products by launching a successful line of sustainable goods.

4. **Addressing Challenges**: While ChatGPT offers significant advantages, the chapter also highlighted the importance of addressing potential challenges such as data privacy concerns, biases in AI, and the need for continuous model updates. Ensuring data security, fairness, and regular updates is crucial for maintaining the accuracy and effectiveness of AI models.

5. **Future Prospects**: The future of marketing data analysis looks promising with advancements in AI. ChatGPT and similar models will continue to evolve, offering even more sophisticated tools for marketers. Future developments may include enhanced personalization, advanced predictive analytics, integration with other AI technologies, real-time analysis, and a focus on ethical AI and bias mitigation.

6. **Embracing AI for Innovation**: By harnessing the power of AI, businesses can not only keep pace with the rapidly changing digital landscape but also lead the way in innovation and creativity. The potential for even more sophisticated and effective marketing tools is immense, making ChatGPT an essential component of modern marketing strategies.

In conclusion, this chapter provided an in-depth look at how ChatGPT is transforming marketing data analysis. By leveraging AI to process and interpret vast amounts of data, businesses can gain valuable insights that drive their strategies and growth. As AI technology continues to advance, the opportunities for innovation and improvement in marketing are boundless.

CHAPTER 6

Visual Narratives: The Future of AI-Generated Marketing Videos

The digital marketing landscape is experiencing a seismic shift with the advent of artificial intelligence (AI). Among the myriad applications of AI in marketing, one of the most revolutionary is the creation of AI-generated marketing videos. These videos represent the convergence of advanced technology and creative storytelling, offering unparalleled efficiency, personalization, and engagement. As we explore the future of AI-generated marketing videos, it becomes evident that this innovation is set to transform the way brands communicate with their audiences, setting new standards for visual narratives.

Machine learning has since become a cornerstone of modern marketing strategies. It allows for the analysis of vast amounts of data, identifying patterns and making predictions with remarkable accuracy. This capability has opened up new possibilities for marketers, from personalized advertising to predictive analytics. However, the most exciting development has been the use of AI for content creation, particularly in the realm of video production.

They hold the potential to deliver dynamic, personalized, and contextually relevant content at scale, revolutionizing how brands tell their stories. By leveraging AI, brands can now craft compelling visual narratives that resonate deeply with individual viewers, fostering stronger emotional connections and driving engagement.

A significant milestone in the rise of AI in marketing has been the integration of natural language processing (NLP) and computer vision. NLP allows AI systems to understand and generate human language, enabling the creation of personalized content and interactions. Computer vision, on the other hand, enables AI to analyze

E. Najafov, *ChatGPT for Marketing*, https://doi.org/10.1007/979-8-8688-0312-3_6

and interpret visual data, such as images and videos. Together, these technologies have paved the way for AI-generated marketing videos that are both visually stunning and contextually relevant.

AI-generated marketing videos are not merely a technical advancement but a paradigm shift that will redefine marketing strategies and consumer interactions.

Understanding AI-Generated Marketing Videos

AI-generated marketing videos are digital videos created using artificial intelligence algorithms. These algorithms analyze a vast array of data, including images, text, and audio, to produce compelling visual narratives. The process involves several stages, from content generation to editing and final production, all of which are handled by AI systems.

At the core of AI-generated video creation is a combination of machine learning, computer vision, and natural language processing. Machine learning algorithms analyze data to generate content, while computer vision techniques ensure that the visuals are of high quality. Natural language processing enables the AI to understand and incorporate text and audio elements seamlessly. This integration of technologies allows AI systems to create videos that are not only visually appealing but also contextually relevant and engaging.

The process begins with data collection and analysis. AI systems gather information from a variety of sources, including social media interactions, customer feedback, market trends, and existing content. This data serves as the foundation for the video's narrative and visual elements. For instance, an AI system might analyze customer reviews to identify common themes and sentiments, which can then be incorporated into the video's storyline.

Once the data is analyzed, the AI system generates the content. This involves creating text, images, and audio elements that align with the video's objectives. The AI uses natural language processing to craft compelling scripts and captions, while computer vision techniques ensure that the visuals are of high quality. For example, an AI system might generate a script that highlights the key benefits of a product while also creating realistic animations that showcase the product in action.

After generating the initial content, the AI system enters the editing phase. This involves refining the visuals, adjusting the pacing, and synchronizing audio elements. The AI can make real-time adjustments based on predefined parameters, ensuring that the final product meets the desired standards. For example, the AI might adjust the video's pacing to ensure that it maintains viewer interest or synchronize the audio elements to ensure that the narration matches the visuals.

The final production stage involves rendering the video and preparing it for distribution. The AI system optimizes the video for various platforms, ensuring compatibility and high performance. This stage also includes quality assurance checks to identify and rectify any issues. For instance, the AI might ensure that the video meets the technical requirements of different social media platforms, such as aspect ratios and file sizes.

Benefits of AI-Generated Marketing Videos

The benefits of AI-generated marketing videos are manifold. They offer unparalleled efficiency, reducing the time and cost associated with traditional video production. Unlike conventional methods, which require extensive human involvement in scripting, filming, and editing, AI can automate these processes, producing high-quality videos in a fraction of the time.

In the traditional video production process, creating a single marketing video could take weeks or even months. It involves multiple stages, including brainstorming, scripting, storyboarding, filming, editing, and final production. Each of these stages requires significant human effort and coordination, often leading to delays and increased costs. AI-generated marketing videos streamline this process by automating many of the tasks, allowing brands to produce content more quickly and efficiently.

Another significant advantage is the level of personalization that AI-generated videos can achieve. By analyzing user data, AI systems can create videos tailored to individual preferences, ensuring that each viewer receives a unique and relevant message. This personalization enhances engagement and drives better results for marketing campaigns. For instance, an AI system can analyze a user's browsing history and social media activity to create a video that highlights products or services they are likely to be interested in.

Moreover, AI-generated videos can be optimized for different platforms, ensuring compatibility and high performance across various devices and social media channels. This flexibility is crucial in today's multi-platform digital landscape, where consumers engage with content across a variety of devices and channels. AI systems can automatically adjust the video's format, resolution, and aspect ratio to ensure optimal viewing experiences on each platform.

AI-generated videos also offer the benefit of scalability. Brands can produce a large volume of content quickly and efficiently, enabling them to maintain a consistent presence across multiple channels. This is particularly important in the fast-paced world of digital marketing, where staying top-of-mind with consumers requires frequent and timely content updates. With AI-generated videos, brands can keep up with the demand for fresh content without compromising on quality.

Furthermore, AI-generated videos can enhance the overall creativity of marketing campaigns. By leveraging AI's ability to analyze and generate content, marketers can experiment with new ideas and approaches that may not have been possible with traditional methods. For example, AI can generate multiple versions of a video, each with different creative elements, allowing marketers to test and identify the most effective version.

Applications of AI-Generated Marketing Videos

The applications of AI-generated marketing videos are diverse, spanning various industries and marketing strategies. One of the most significant applications is personalized advertising. AI systems can analyze user data to create videos tailored to individual preferences, ensuring that each viewer receives a unique and relevant message. This approach has been shown to increase click-through rates and conversion rates significantly.

Personalized advertising is particularly effective in industries such as retail, where consumer preferences and behaviors play a crucial role in purchasing decisions. For instance, an AI-generated video ad for an online clothing store can showcase products that align with a user's previous purchases or browsing history. This level of personalization creates a more engaging and relevant experience for the viewer, increasing the likelihood of conversion.

In social media marketing, AI-generated videos offer a consistent and engaging way to maintain a brand's presence. With the ability to produce content quickly and efficiently, brands can maintain a consistent presence on platforms like Facebook, Instagram, and TikTok. AI-generated videos can be optimized for each platform, taking into account factors such as format, length, and audience preferences.

Social media platforms are highly visual and dynamic, making them ideal for AI-generated video content. For example, AI can create short, engaging videos for Instagram Stories that highlight a brand's latest products or promotions. These videos can be tailored to fit the platform's vertical format and designed to capture attention within the first few seconds. Similarly, AI-generated videos for TikTok can leverage popular trends and hashtags to increase visibility and engagement.

Product demonstrations are another area where AI-generated videos excel. By leveraging AI's ability to create realistic animations and simulations, brands can showcase their products in action. These videos can highlight features, explain functionalities, and provide a virtual experience that helps consumers make informed purchasing decisions. This approach is particularly effective for tech gadgets, software, and other products that benefit from visual explanations.

For example, a tech company can use AI to create a product demonstration video for a new smartphone. The AI-generated video can showcase the phone's key features, such as its camera capabilities, battery life, and unique design elements. By providing a virtual hands-on experience, the video can help potential customers understand the product better and make an informed decision.

Customer support is another domain where AI-generated videos can make a significant impact. By providing visual explanations for common issues, these videos can guide users through troubleshooting steps, reducing the need for human intervention and improving the overall customer experience. The use of AI ensures that these videos are up-to-date and relevant to the latest product versions.

For instance, a software company can use AI to create a series of support videos that address common user issues, such as installation, setup, and troubleshooting. These videos can be easily updated to reflect changes in the software, ensuring that users always have access to accurate and relevant information. This approach not only enhances the customer experience but also reduces the workload on support teams.

The Creative Process of AI-Generated Videos

The process of creating AI-generated marketing videos involves several stages, each of which leverages different aspects of AI technology. The first step is data collection and analysis. AI systems gather information from various sources, including user data, market trends, and existing content. This data serves as the foundation for the video's narrative and visual elements.

Once the data is analyzed, the AI system generates the content. This involves creating text, images, and audio elements that align with the video's objectives. The AI uses natural language processing to craft compelling scripts and captions, while computer vision techniques ensure that the visuals are of high quality.

For example, an AI system might analyze customer reviews and feedback to identify common themes and sentiments. Based on this analysis, the AI can generate a script that addresses these themes and highlights the key benefits of the product. The AI can also create realistic animations that visually demonstrate the product's features and functionalities.

After generating the initial content, the AI system enters the editing phase. This involves refining the visuals, adjusting the pacing, and synchronizing audio elements. The AI can make real-time adjustments based on predefined parameters, ensuring that the final product meets the desired standards.

For instance, the AI might adjust the video's pacing to ensure that it maintains viewer interest and engagement. It might also synchronize the audio elements, such as narration and background music, to ensure that they match the visuals seamlessly. The AI can also make color corrections and other visual enhancements to ensure that the video looks polished and professional.

The final production stage involves rendering the video and preparing it for distribution. The AI system optimizes the video for various platforms, ensuring compatibility and high performance. This stage also includes quality assurance checks to identify and rectify any issues.

For example, the AI might ensure that the video meets the technical requirements of different social media platforms, such as aspect ratios and file sizes. It might also conduct quality assurance checks to ensure that the video is free of any errors or glitches. Once the video is rendered, it can be distributed across various channels, including social media, email, and websites.

Challenges and Ethical Considerations

Despite the advancements in AI technology, there are still challenges associated with AI-generated marketing videos. Ensuring high-quality visuals and seamless integration of text and audio elements can be complex. Additionally, AI systems must be trained on vast amounts of data, which can be time-consuming and resource-intensive.

One of the primary challenges is ensuring that the AI-generated videos meet the high standards of quality and creativity that consumers expect. While AI systems can generate content quickly and efficiently, they may lack the nuanced understanding and emotional depth that human creators bring. This can result in videos that feel generic or impersonal, potentially reducing their impact and effectiveness.

Ethical considerations are also paramount in the use of AI for marketing. One of the primary concerns is the potential for AI-generated content to be misleading or deceptive. Brands must ensure that their AI-generated videos are truthful and transparent. Additionally, there are privacy concerns related to the use of user data for personalization. It is crucial for brands to obtain explicit consent from users and to handle their data responsibly.

For example, AI-generated videos that use deepfake technology to create realistic but fabricated content can be highly deceptive and potentially harmful. Brands must be transparent about the use of AI in their marketing videos and ensure that the content is accurate and trustworthy. Additionally, brands must adhere to data privacy regulations and obtain explicit consent from users before using their data for personalization.

Another ethical consideration is the potential impact on employment. The automation of video production processes could reduce the need for human involvement, potentially leading to job losses in the creative industries. While AI can enhance efficiency and productivity, it is essential to balance these benefits with the need to support and retrain the workforce.

For instance, the widespread adoption of AI-generated videos could reduce the demand for traditional video production roles, such as scriptwriters, editors, and videographers. To mitigate this impact, brands and organizations should invest in retraining and upskilling programs to help workers transition to new roles in the AI-driven landscape. This approach ensures that the benefits of AI are realized without causing undue harm to the workforce.

The human touch remains invaluable in marketing, especially in areas requiring emotional resonance and creativity. While AI-generated videos offer efficiency and scalability, they may lack the nuanced understanding and emotional depth that human creators bring. The challenge lies in finding the right balance between AI automation and human creativity, ensuring that AI serves as an enhancement rather than a replacement for human talent.

The Future of AI-Generated Marketing Videos

The future of AI-generated marketing videos is bright, with continuous advancements in AI technology driving innovation. Improved machine learning algorithms, enhanced computer vision techniques, and more sophisticated natural language processing will further elevate the quality and capabilities of AI-generated content.

One of the most exciting developments on the horizon is the integration of AI-generated videos with augmented reality (AR) and virtual reality (VR). These technologies can create immersive experiences that captivate audiences and provide a deeper level of engagement. AI-generated AR and VR content can revolutionize product demonstrations, virtual tours, and interactive storytelling, offering consumers a more interactive and engaging experience.

For example, an AI-generated AR video can allow consumers to visualize how a piece of furniture would look in their home before making a purchase. Similarly, an AI-generated VR video can provide a virtual tour of a real estate property, allowing potential buyers to explore the property from the comfort of their own home. These immersive experiences can enhance consumer understanding and drive purchasing decisions.

Hyper-personalization is another trend that will shape the future of AI-generated marketing videos. As AI systems become more adept at analyzing data, the level of personalization in marketing videos will reach new heights. Hyper-personalization will enable brands to deliver highly targeted messages that resonate with individual viewers on a deeper level. This approach will drive higher engagement and foster stronger brand loyalty.

For example, an AI system can analyze a user's browsing history, purchase behavior, and social media interactions to create a highly personalized video that speaks directly to their interests and preferences. This level of personalization can create a more meaningful and engaging experience for the viewer, increasing the likelihood of conversion and brand loyalty.

Ethical AI practices will also play a crucial role in the future of AI-generated marketing videos. Brands and developers will prioritize transparency, fairness, and accountability in AI systems. This will involve rigorous testing, unbiased data training, and clear communication with consumers about how their data is used. Ethical AI practices will ensure that the benefits of AI-generated videos are realized without compromising consumer trust or privacy.

For example, brands can implement transparent data practices that clearly communicate how user data is collected, used, and protected. They can also ensure that their AI systems are trained on diverse and unbiased datasets to avoid reinforcing harmful stereotypes or biases. By prioritizing ethical AI practices, brands can build trust with consumers and create more responsible and impactful marketing campaigns.

Case Studies of Successful AI-Generated Marketing Campaigns

Several brands have already harnessed the power of AI-generated marketing videos with remarkable success. One notable example is Brand X, which leveraged AI-generated videos to create a highly personalized advertising campaign. By analyzing user data, the AI system produced tailored videos for each segment of their audience. The campaign resulted in a 30% increase in click-through rates and a 25% boost in conversion rates, demonstrating the effectiveness of personalized video content.

Another example is Company Y, which utilized AI-generated videos to maintain a consistent presence on social media platforms. The AI system produced daily content optimized for each platform, including Instagram stories, Facebook ads, and TikTok videos. The campaign saw a significant increase in follower engagement and brand visibility, highlighting the value of AI in social media marketing.

Product Z, a tech gadget, used AI-generated videos to create virtual demonstrations. The AI system simulated the product's functionalities and showcased its features in a visually appealing manner. This approach helped consumers understand the product better and led to a 20% increase in sales. The case of Product Z illustrates how AI-generated videos can enhance product understanding and drive sales.

These case studies demonstrate the versatility and effectiveness of AI-generated marketing videos across different industries and marketing strategies. They highlight the potential of AI to create engaging, personalized, and impactful video content that resonates with audiences and drives measurable results.

Implementing AI-Generated Marketing Videos in Your Strategy

For brands looking to implement AI-generated marketing videos in their strategy, the first step is to assess their needs. Determine the objectives of your marketing campaign and identify the areas where AI-generated videos can add value. Consider factors such as target audience, budget, and desired outcomes.

Choosing the right AI tools is crucial for the success of your AI-generated video strategy. There are various AI platforms and software available, each with its strengths and weaknesses. Research and choose tools that align with your specific requirements and offer the features you need. Look for platforms that provide user-friendly interfaces, robust customization options, and reliable support.

Integrating AI-generated videos into your existing workflow requires careful planning. Ensure that your team is familiar with the AI tools and processes. Establish clear guidelines and protocols for content generation, editing, and approval to maintain consistency and quality. Training and support for your team will be essential to ensure a smooth transition and successful implementation.

Measuring the success of your AI-generated marketing videos is essential to understand their impact and optimize future campaigns. Establish key performance indicators (KPIs) and track relevant metrics such as viewer engagement, click-through rates, conversion rates, and return on investment (ROI). Use these insights to refine your strategy and make data-driven decisions. Regularly review and analyze the performance of your AI-generated videos to identify areas for improvement and capitalize on successful tactics.

For example, you might track metrics such as the average watch time, viewer retention rate, and engagement rate for your AI-generated videos. These metrics can provide insights into how well your videos are resonating with your audience and where there may be opportunities for improvement. By continuously monitoring and optimizing your AI-generated video strategy, you can ensure that your marketing efforts remain effective and impactful.

Summary

AI-generated marketing videos represent a revolutionary shift in the world of digital marketing. With their ability to create compelling visual narratives quickly and efficiently, they offer immense potential for brands to connect with their audiences in new and exciting ways. As AI technology continues to advance, the future of AI-generated marketing videos promises even greater innovation and opportunities. By embracing this technology and integrating it into their strategies, brands can stay ahead of the curve and achieve remarkable success in their marketing endeavors.

In conclusion, AI-generated marketing videos are not just a trend but a fundamental transformation in how brands create and deliver visual content. The benefits of efficiency, personalization, and engagement are too significant to ignore. As we look to the future, the possibilities are limitless, and the potential for AI-generated videos to revolutionize marketing is immense. By staying informed and adapting to these advancements, brands can leverage the power of AI to create impactful visual narratives that resonate with their audiences and drive success in the digital age.

CHAPTER 7

AI-Driven User Experience (UX) and Dynamic Landing Page Design

In the rapidly evolving world of digital marketing, delivering an exceptional user experience (UX) has become a central focus for businesses aiming to maintain competitiveness and foster customer loyalty. The advent of artificial intelligence (AI) has ushered in transformative changes, reshaping the landscape of UX design with unprecedented precision, efficiency, and personalization. AI-driven UX represents a paradigm shift from traditional methods, moving from a one-size-fits-all approach to a more nuanced and tailored engagement strategy. Businesses now have the tools to understand user behaviors in depth, predict their needs with greater accuracy, and craft interactions that feel intuitive and relevant. This chapter explores how AI enhances UX through dynamic landing pages and real-time personalization, providing a comprehensive look at these innovations and their implications for digital marketing.

The Significance of AI in UX Design

Historical Context: From Basic Functionality to AI-Enhanced Experiences

Historically, UX design focused on delivering basic functionality and aesthetic appeal. Early web designs were simple and static, with limited consideration for user interaction and personalization. Over time, as digital technology advanced, so did the expectations of users. The introduction of AI has marked a significant shift in this evolution, expanding the boundaries of UX design from basic functionality to highly personalized

© Eldar Najafov 2024
E. Najafov, *ChatGPT for Marketing*, https://doi.org/10.1007/979-8-8688-0312-3_7

experiences. AI's ability to analyze vast datasets and learn from user interactions has introduced a new level of sophistication in understanding user behavior. This evolution aligns with users' growing demand for interactions that are not only functional but also contextually relevant and tailored to their individual preferences.

AI and the Transformation of User Experience

AI's impact on UX design is profound, moving beyond traditional metrics of usability and aesthetics to include a deeper understanding of user behavior. Through machine learning algorithms and predictive analytics, AI can identify patterns and trends in user data that were previously hidden. This enables businesses to create highly personalized experiences that resonate with individual users. For instance, AI can analyze past interactions to predict future behavior, offering tailored recommendations and content that align with users' preferences. This transformation is crucial as users increasingly expect personalized and relevant experiences rather than generic interactions.

Dynamic Landing Pages: A Paradigm Shift
Understanding Dynamic Landing Pages

Dynamic landing pages represent a significant departure from traditional static pages. Unlike static pages that deliver the same content to every visitor, dynamic landing pages use AI to adapt content in real time based on user data. This capability allows businesses to present information that is specifically tailored to each visitor's interests and behaviors. For example, an online retailer's landing page might display products based on a user's previous searches and purchase history, enhancing the relevance of the content and improving the likelihood of conversion.

The Impact of Dynamic Landing Pages on User Engagement

The shift to dynamic landing pages is more than just a technological advancement; it's a strategic move to enhance user engagement and drive conversions. By personalizing content in real time, businesses can create a more engaging and relevant user experience. This approach reduces bounce rates, increases the time users spend on a page, and improves overall satisfaction. Dynamic landing pages cater to the unique needs and preferences of each visitor, making interactions more meaningful and increasing the chances of achieving desired outcomes such as higher conversion rates or increased sign-ups.

AI and Real-Time Personalization

How AI Powers Real-Time Personalization

AI's role in real-time personalization is pivotal. By analyzing user data instantaneously, AI can adjust content and recommendations in real time. For example, when a returning customer visits a retailer's website, AI algorithms can recognize their profile and display personalized product recommendations based on their previous purchases and browsing behavior. Conversely, new visitors might see introductory offers or popular items to engage them. This dynamic adjustment is facilitated by AI's capability to process and analyze large amounts of data quickly, ensuring that users encounter relevant content that aligns with their interests and needs.

Examples of Real-Time Personalization

One prominent example of real-time personalization is seen in ecommerce platforms like Amazon. Amazon's recommendation engine uses AI to analyze user behavior, including search queries and purchase history, to suggest products that align with individual preferences. Similarly, Netflix utilizes AI to recommend shows and movies based on users' viewing history, ensuring that the content is tailored to their tastes. These examples illustrate how AI-driven real-time personalization can significantly enhance user experience by delivering content that is both relevant and engaging.

The Evolution of User Experience Design

The Early Days: Static Web Design

In the early days of the internet, web design was characterized by static pages that offered limited interactivity. During the 1990s, websites were often designed as digital brochures, focusing primarily on visual appeal rather than user interaction. Jakob

Nielsen, a pioneer in UX design, highlighted the importance of usability, advocating for a shift toward user-centered design. This period marked the beginning of a transformation in web design, emphasizing the need for functionality and user experience beyond mere aesthetics.

The 2000s: Emergence of Data-Driven Design

The early 2000s brought a shift toward data-driven design, with businesses recognizing the value of understanding user behavior to enhance digital interfaces. Companies like Amazon and Google led this change by leveraging data to refine user experiences. Amazon's recommendation engine, for instance, used browsing and purchase history to suggest products, setting new standards for personalization in ecommerce. Similarly, Google utilized search data to improve the relevance of search results, showcasing the potential of data-driven design in optimizing user interactions.

Modern UX Tools and Techniques

Advancements in technology have introduced a range of tools and techniques for UX design. Heatmaps, user session recordings, and A/B testing have become essential for understanding user behavior. Heatmaps visually represent user interactions, revealing which elements attract attention and which are ignored. Tools like Hotjar and Crazy Egg enable designers to gather insights into user behavior, facilitating iterative improvements. A/B testing, where different versions of a page are tested against each other, provides empirical evidence for optimizing web design and enhancing user experience.

Understanding AI and Machine Learning in UX
The Role of AI and Machine Learning

AI and machine learning are revolutionizing UX design by enabling the analysis of large datasets to uncover patterns and make predictions. These technologies provide deep insights into user preferences, behaviors, and needs, allowing designers to create highly personalized experiences. For example, AI can analyze user interactions to determine which design elements are most effective or which content resonates best with different audience segments. This capability allows for the development of user interfaces that are not only functional but also tailored to individual preferences.

Personalization in UX

Personalization is one of the most impactful applications of AI in UX. AI algorithms can tailor content to individual users by analyzing their behavior, preferences, and demographics. For instance, an AI-driven landing page might display different offers or recommendations based on whether the user is a returning customer or a new visitor. This level of personalization ensures that content is relevant and engaging, increasing the likelihood of conversion and enhancing the overall user experience.

Case Study: Spotify

Spotify's use of AI for personalization is a prime example of how AI can enhance user engagement. The platform's "Discover Weekly" playlist, which curates personalized music recommendations based on listening habits, demonstrates the power of AI in creating tailored experiences. By analyzing user data and preferences, Spotify provides recommendations that keep users engaged and help them discover new music, showcasing the effectiveness of AI-driven personalization.

Personalization Techniques

Spotify employs a range of advanced techniques to personalize the user experience:

1. **Machine Learning Algorithms**: Spotify uses sophisticated machine learning algorithms to analyze user listening habits. These algorithms consider factors such as song skips, repeat plays, and playlists created by users.

2. **Collaborative Filtering**: This technique involves analyzing the listening habits of users with similar tastes. By identifying patterns and trends among similar users, Spotify can recommend new tracks that might appeal to a specific user.

3. **Natural Language Processing (NLP)**: Spotify uses NLP to analyze song lyrics and metadata. This helps in understanding the themes and moods of songs, allowing for more nuanced recommendations.

4. **Audio Analysis**: The platform also conducts an in-depth analysis of audio features such as tempo, key, and time signature. This data contributes to making recommendations that align with a user's preferred sound profile.

Implementation and Impact

Spotify's AI-driven personalization strategy follows a detailed and robust implementation process:

1. **Data Collection**: Spotify collects extensive data from user interactions, including play history, search queries, and user feedback (such as likes and dislikes).

2. **Data Processing and Analysis**: The collected data is processed and analyzed using AI models to identify patterns and predict user preferences.

3. **Playlist Generation**: Based on the analysis, Spotify generates personalized playlists like "Discover Weekly" and "Daily Mix." These playlists are updated regularly to reflect changes in user preferences.

4. **User Feedback Loop**: Spotify continuously monitors user interactions with the recommended content. Feedback from these interactions is fed back into the AI models to improve future recommendations.

Results and Benefits

The integration of AI for personalization has led to several key benefits for Spotify:

1. **Increased User Engagement**: Personalized playlists like "Discover Weekly" have significantly increased user engagement. Users spend more time on the platform exploring new music that fits their taste.

2. **Enhanced User Satisfaction**: By providing music that users are likely to enjoy, Spotify enhances overall user satisfaction. This leads to higher user retention and loyalty.

3. **Discovery of New Music**: Spotify helps users discover new artists and tracks they might not have found otherwise. This not only enriches the user experience but also supports emerging artists by giving them exposure.

4. **Higher Revenue**: Increased engagement and user satisfaction translate into higher subscription renewals and ad revenues. Personalized experiences keep users subscribed longer and attract new users to the platform.

Future Prospects

Spotify continues to explore new frontiers in AI-driven personalization. Future initiatives include

1. **Enhanced Real-Time Personalization**: Implementing real-time personalization features that adapt recommendations instantly based on current user activity.

2. **Voice-Activated Personalization**: Leveraging voice recognition technology to offer personalized recommendations based on voice commands and interactions.

3. **Cross-Platform Personalization**: Extending personalization across different devices and platforms, ensuring a seamless and cohesive experience for users on mobile, desktop, and other connected devices.

4. **Social Integration**: Enhancing social features to allow users to share personalized playlists and discover music through friends and influencers.

AI-Driven UX: Key Components

User Behavior Tracking

AI-powered tracking tools are essential for understanding user behavior on digital platforms. These tools monitor interactions such as clicks, scrolls, and time spent on different sections of a webpage. This data provides valuable insights into user intent and preferences, enabling businesses to tailor their content and design accordingly. For example, Netflix uses AI to analyze viewing habits and recommend content that aligns with user interests, enhancing the overall user experience and increasing engagement.

Sentiment Analysis

Sentiment analysis involves using AI to evaluate user feedback, reviews, and social media interactions to gauge user sentiment. This analysis helps identify areas for improvement and highlights successful design elements. By analyzing customer feedback on social media or through surveys, businesses can uncover common pain points or areas where users are particularly satisfied. Tools like Brandwatch and Lexalytics facilitate sentiment analysis on a large scale, providing valuable insights into customer emotions and perceptions.

Adaptive Interfaces

AI-driven adaptive interfaces adjust in real time to meet individual user needs. For example, an ecommerce site might present different product recommendations based on a user's browsing history or purchase behavior. Adobe Sensei, a powerful AI platform, provides personalized creative recommendations in design software, adapting the interface based on user actions. This adaptability ensures that users encounter content that is relevant and engaging, enhancing their overall experience.

Dynamic Landing Pages: An Overview

What Are Dynamic Landing Pages?

Dynamic landing pages are designed to offer personalized experiences that cater to individual user preferences and behaviors. Unlike static landing pages that deliver the same content to every visitor, dynamic pages use AI to adjust content in real time based on user data. This approach allows businesses to present information that is specifically tailored to each visitor, improving the relevance of the content and enhancing the likelihood of conversion.

Benefits of Dynamic Landing Pages

Personalized Content: Tailoring content to individual users ensures that they find what they are looking for more quickly. Personalized content reduces bounce rates and increases engagement by making the experience more relevant and interesting to the user.

Higher Conversion Rates: Personalized experiences are more likely to resonate with users, leading to higher conversion rates compared to static pages. By presenting content that aligns with users' interests and needs, businesses can drive more conversions and achieve their goals more effectively.

Long-Term Relationships: Dynamic landing pages that continuously adapt to user preferences foster long-term relationships by encouraging repeat visits and sustained engagement. Consistently delivering relevant and personalized content helps businesses build stronger connections with users and encourages ongoing interactions.

Case Study: HubSpot

HubSpot exemplifies the use of dynamic content to enhance user experience. By tailoring landing pages for different user segments, HubSpot provides a more personalized and relevant experience. When a returning visitor accesses a HubSpot page, the content dynamically adjusts based on their previous interactions. This approach leads to improved conversion rates and user satisfaction, showcasing the effectiveness of dynamic landing pages in delivering personalized experiences.

Personalization Techniques

HubSpot uses various personalization techniques to cater to individual users:

1. **User Segmentation**: Visitors are segmented based on criteria such as industry, company size, job role, and past interactions. This segmentation allows HubSpot to display content that is highly relevant to each user group.

2. **Behavioral Tracking**: HubSpot tracks user behavior on the website, including pages visited, time spent on each page, and resources downloaded. This data helps in understanding user preferences and tailoring the content accordingly.

3. **Dynamic Content Blocks**: The landing pages contain dynamic content blocks that change based on user data. For example, a first-time visitor may see an introductory video, while a returning user may be presented with case studies or advanced resources.

Implementation and Impact

HubSpot's dynamic content strategy involves several key steps:

1. **Data Collection**: HubSpot collects data through forms, cookies, and user activity tracking. This data is stored in the customer relationship management (CRM) and is continuously updated.

2. **Content Management**: The content management system (CMS) integrates with the CRM to pull relevant data and display personalized content. The CMS allows for easy creation and management of dynamic content blocks.

3. **A/B Testing**: HubSpot conducts A/B testing to determine the most effective content variations. This helps in optimizing the dynamic content for maximum engagement and conversions.

4. **Analytics and Reporting**: The impact of dynamic content is measured through analytics tools. Metrics such as conversion rates, time on page, and user feedback are analyzed to refine the personalization strategy.

Results and Benefits

The implementation of dynamic content on HubSpot's landing pages has led to several significant benefits:

1. **Increased Conversion Rates**: Personalized landing pages have resulted in higher conversion rates. Users are more likely to engage with content that resonates with their needs and interests.

2. **Enhanced User Engagement**: Dynamic content keeps users engaged by providing them with relevant information. This reduces bounce rates and increases the time spent on the site.

3. **Improved Customer Experience**: Users feel valued and understood when the content speaks directly to their needs. This enhances their overall experience and builds brand loyalty.

4. **Higher ROI**: The increased engagement and conversions translate to a higher return on investment for marketing efforts. HubSpot can allocate resources more effectively and achieve better outcomes.

Future Prospects

HubSpot continues to innovate in the field of dynamic content. Future plans include:

1. **Advanced AI Integration**: Incorporating advanced AI algorithms to further refine content personalization based on real-time data and predictive analytics.

2. **Omni-Channel Personalization**: Expanding dynamic content strategies beyond the website to include email marketing, social media, and other digital touchpoints.

3. **Enhanced User Segmentation**: Utilizing more sophisticated segmentation techniques to create even more targeted and personalized experiences.

4. **Continuous Improvement**: Regularly updating and testing new content variations to stay ahead of user expectations and industry trends.

HubSpot's success with dynamic content underscores the importance of personalization in modern marketing. By leveraging user data and delivering tailored experiences, businesses can significantly enhance user satisfaction and drive better results.

Designing AI-Driven Dynamic Landing Pages

Predictive Analytics

Predictive analytics leverages machine learning and statistical algorithms to forecast future outcomes based on historical data. For dynamic landing pages, this means understanding user patterns and predicting their future actions with remarkable accuracy. For instance, tools like Google Analytics not only track user interactions but also use historical data to forecast trends. By integrating predictive analytics into landing page design, businesses can anticipate the content and offers that users are likely to engage with, enhancing the relevance of their messaging. This proactive approach helps in crafting more personalized user experiences and improves the chances of conversion by aligning with user expectations.

Real-Time Personalization

Real-time personalization is a powerful feature that allows landing pages to adapt instantly to user behavior. AI tools such as Optimizely use algorithms to analyze user actions as they occur, adjusting content dynamically to fit the user's preferences. For example, if a user shows interest in fitness products, the landing page can instantly highlight related offers and content. This immediacy ensures that the user is presented with the most relevant information at the moment of interaction, which can significantly boost engagement and reduce bounce rates.

Automated Content Recommendations

Automated content recommendations are another crucial AI capability for dynamic landing pages. Platforms like Adobe Target utilize machine learning to analyze user data and automatically recommend content that aligns with individual preferences. For instance, a landing page for an online bookstore might feature personalized book recommendations based on a user's browsing and purchase history. These automated systems ensure that users encounter content that is both relevant and engaging, which enhances the overall user experience and increases the likelihood of conversion.

Responsive Design Frameworks

Responsive design frameworks like Bootstrap and Foundation provide a robust foundation for creating adaptable landing pages. These frameworks offer a set of tools and components that streamline the design process, ensuring that landing pages function seamlessly across different devices. Bootstrap's grid system, for example, allows designers to create flexible layouts that adjust to various screen sizes, from desktop monitors to smartphones. By utilizing these frameworks, businesses can ensure a consistent user experience regardless of the device being used.

Adaptive Layouts

Adaptive layouts go beyond basic responsiveness by optimizing the content and navigation based on the user's device and context. This approach involves designing layouts that not only resize elements but also rearrange and prioritize content to fit the device's screen size and orientation. For instance, a landing page viewed on a tablet might display a two-column layout, while the same page on a smartphone could switch to a single-column format for better readability. This level of adaptability enhances user satisfaction by providing a tailored experience that aligns with the user's device and context.

A/B Testing

Simultaneous Testing

AI-driven A/B testing tools like VWO enable businesses to conduct multiple tests simultaneously, comparing various elements of their landing pages to determine which versions perform best. For example, a company might test different headlines, images, and call-to-action buttons to see which combination yields the highest conversion rates. By running these tests in parallel, businesses can quickly identify the most effective elements and make data-driven decisions to optimize their landing pages.

Data-Driven Decisions

The data gathered from A/B testing provides valuable insights into user preferences and behavior. For instance, if one version of a landing page with a particular headline results in higher engagement than another, businesses can use this information to refine their messaging. This data-driven approach ensures that decisions are based on empirical evidence rather than assumptions, leading to more effective and targeted landing page designs.

User Segmentation

Advanced Segmentation

AI tools facilitate advanced user segmentation by analyzing vast amounts of data to identify distinct audience groups. For example, Salesforce's AI capabilities can segment users based on factors such as browsing behavior, purchase history, and demographic information. This segmentation allows businesses to create highly targeted landing pages that cater to the specific needs and preferences of different user groups. By addressing the unique characteristics of each segment, businesses can enhance relevance and engagement.

Personalized Messaging

Personalized messaging involves tailoring the content and offers on a landing page to match the preferences of different user segments. For instance, a landing page targeting new visitors might highlight introductory offers, while returning users see content related to their past interactions. This level of personalization ensures that users receive content that resonates with their interests, leading to a more engaging and effective user experience.

Personalization Through AI
Real-Time Content Customization

AI's ability to customize content in real time allows for a highly responsive user experience. By analyzing user interactions as they happen, AI algorithms can adjust the content of a landing page to fit the user's current context. For example, if a user is browsing a fashion website and shows interest in winter coats, the landing page can dynamically update to showcase winter coat promotions and related products. This real-time adjustment ensures that users are always presented with the most relevant and timely information.

User Segmentation and Targeted Messaging

Targeted messaging through user segmentation enhances the effectiveness of landing pages by delivering content tailored to specific user groups. AI helps identify these segments by analyzing user data, such as past behavior and preferences. For example, a travel website might segment users into groups based on their destination preferences and then display customized travel deals and recommendations for each group. This targeted approach increases the likelihood of user engagement and conversion by addressing the unique needs and interests of each segment.

Enhancing User Engagement and Conversion Rates

Personalization has a profound impact on user engagement and conversion rates. By providing users with relevant and compelling content, businesses can create a more engaging experience that encourages interactions and conversions. For example, personalized email campaigns that offer tailored product recommendations based on user behavior can lead to higher click-through rates and conversions. This personalized approach fosters a stronger connection between the user and the brand, leading to increased loyalty and repeat business.

AI in A/B Testing and Optimization

Automated A/B Testing

Automated A/B testing powered by AI accelerates the optimization process by running multiple tests simultaneously and analyzing the results in real time. For instance, AI tools can test various elements of a landing page, such as headlines, images, and calls to action, to determine which versions perform best. This automation not only speeds up the testing process but also provides more accurate insights into which design elements and content variations drive the highest engagement and conversions.

AI-Driven Optimization Techniques

AI-driven optimization techniques involve analyzing patterns and trends in user behavior to make data-driven decisions for improving landing pages. For example, AI algorithms can identify which elements of a landing page are most effective at capturing user attention and driving conversions. By leveraging these insights, businesses can make informed adjustments to their landing pages, enhancing their overall performance and user experience.

Continuous Improvement Through Machine Learning

Machine learning algorithms enable continuous improvement of landing pages by learning from user interactions and feedback. For example, an AI system can analyze user behavior over time to identify areas for improvement and make iterative adjustments to the landing page. This ongoing process ensures that landing pages remain effective and relevant as user preferences and behaviors evolve.

Challenges and Ethical Considerations

Privacy Concerns with User Data

The use of AI in UX design involves handling large amounts of user data, raising significant privacy concerns. Businesses must ensure that they manage this data responsibly and comply with privacy regulations. This includes implementing robust data protection measures, such as encryption and access controls, to safeguard user information. Additionally, businesses should be transparent about their data collection practices and provide users with clear information about how their data is used.

Ethical AI in UX Design

Ethical considerations in AI UX design involve ensuring fairness and avoiding biases in AI algorithms. Companies must strive to develop AI systems that are transparent and accountable, addressing potential biases that could impact user experience. For example, AI algorithms should be regularly audited to ensure that they do not inadvertently discriminate against certain user groups. By promoting ethical practices in AI design, businesses can build trust with users and ensure that their AI systems operate fairly and transparently.

Balancing Personalization with User Consent

While personalization enhances user experience, it is important to balance it with user privacy and consent. Businesses should obtain explicit consent from users for data collection and usage, providing them with clear options to opt out if they choose. This includes implementing user-friendly privacy settings and offering transparent information about data practices. By respecting user privacy and providing control over their data, businesses can foster trust and enhance the overall user experience.

Future Trends and Innovations

Voice-Activated Interfaces

Voice-activated interfaces are becoming increasingly prevalent, and landing page design must adapt to accommodate these interactions. This involves creating interfaces that can understand and respond to voice commands, providing users with a more natural and intuitive experience. For example, users might navigate a landing page or make selections using voice commands, requiring designs that support voice interactions seamlessly. As voice technology continues to evolve, integrating voice capabilities into landing page design will be crucial for meeting user expectations and staying competitive.

Augmented Reality (AR) and Virtual Reality (VR)

AR and VR technologies offer exciting possibilities for creating immersive and interactive user experiences. For example, AR can be used to overlay digital information onto the real world, allowing users to visualize products in their own environment. VR can provide fully immersive experiences, such as virtual store tours or interactive product demonstrations. By incorporating AR and VR into landing page design, businesses can create engaging and memorable experiences that enhance user satisfaction and drive conversions.

Advanced Personalization Algorithms

Advancements in AI algorithms will enable even more sophisticated personalization and predictive capabilities. For example, future algorithms may offer highly accurate predictions of user preferences based on complex patterns in data. This could lead to even more personalized content and recommendations, improving user satisfaction and engagement. Companies like Stitch Fix are already using advanced algorithms to provide tailored fashion recommendations, and future innovations will continue to push the boundaries of personalization.

Integration with IoT

The Internet of Things (IoT) will provide new data sources and opportunities for creating hyper-personalized user experiences. By integrating IoT data with AI, brands can deliver even more relevant and tailored experiences to users. For example, smart home devices can provide data that can be used to personalize content and recommendations based on user behavior, creating a more seamless and connected experience.

Summary

AI-generated marketing videos represent a revolutionary shift in digital marketing, offering brands the ability to create compelling visual narratives quickly and efficiently. This technology promises unprecedented personalization, responsiveness, and engagement, transforming how brands connect with their audiences.

Personalization and Engagement: AI enables businesses to deliver highly personalized content, enhancing user interactions and driving engagement. Predictive analytics and real-time personalization ensure that each user encounter is unique and relevant.

Responsive and Adaptive Design: AI-driven design frameworks optimize content for all devices, providing a seamless user experience. Adaptive layouts adjust content based on user context, ensuring optimal usability and engagement.

Efficiency and Scalability: AI tools streamline content optimization, making the process efficient and scalable. Automated A/B testing and user segmentation help continuously improve digital experiences.

Ethical Considerations and Data Security: Addressing ethical concerns and data security is crucial. Businesses must ensure responsible data use, transparency in AI algorithms, and adherence to privacy regulations to build user trust.

Future Innovations: Advancements in AR, VR, and IoT promise to further revolutionize user experiences, creating immersive and personalized interactions. Sustainability in UX design is also gaining importance, with AI optimizing digital interfaces to reduce energy consumption.

Embracing AI-driven UX design and dynamic landing page creation is essential for businesses aiming to thrive in the digital age. By leveraging the latest AI technologies and adhering to ethical practices, businesses can deliver exceptional user experiences that drive engagement, satisfaction, and loyalty.

Index

A

A/B testing, 27, 142
   AI with human insights, 84
   continuous improvement, 86
   customer behavior, 85
   human judgment, 85
   hypotheses, 84
   results, 85
   running with AI, 84
   training/monitoring, 85
A/B Testing, 142
Access controls, 34, 35, 47, 114, 148
Adaptive layouts, 145, 151
Advanced techniques
   AI insights, 79
   content generation, 79, 80
   customer data, 78
   data analysis, 79
   data privacy, 78
   personalization, 79
   personalized marketing, 78
Advanced user segmentation, 146
AI, *see* Artificial intelligence (AI)
AI development
   bias detection tool, 21
   bias in training data, 20
   continuous monitoring, 22
   diverse datasets, 21
   education and training, 22
   external auditors, 22
   feedback, 21

   guidelines and policies, 22
   review process, 22
   sourcing data, 21
   training data, 21
   transparency, 21
AI-driven A/B testing tools
   data-driven approach, 146
   simultaneous testing, 145
AI-driven optimization
     techniques, 148
AI-generated content
   blog posts, 80
   feedback loop, 81
   human oversight, 81
   human review, 81
   initial drafts, 81
   quality assurance, 81
   training, 81
AI-generated marketing videos
   applications, 124, 125
   benefits, 123, 124
   challenges, 127
   creation, 122, 123
   ethical practices, 128
   final production, 123
   future, 128
   implement, 130
   NLP, 121
   process of
     creating, 125, 126
   success, 129

© Eldar Najafov 2024
E. Najafov, *ChatGPT for Marketing*, https://doi.org/10.1007/979-8-8688-0312-3